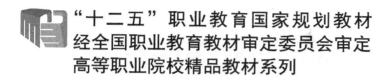

"十二五"职业教育国家规划教材
经全国职业教育教材审定委员会审定
高等职业院校精品教材系列

机械制造类专业英语

杨 成 主编

电子工业出版社
Publishing House of Electronics Industry
北京·BEIJING

内 容 简 介

本书按照教育部最新的职业教育改革要求，结合国家示范性院校专业建设成果及中澳职业教育合作项目成果，充分听取专业教师和行业专家意见进行编写。本书主要介绍机械设计与制造、数控技术、模具设计与制造等专业有关的职场英语知识。全书内容均选自英、美等国专业教材和专业刊物中的原文。为力求用语地道，对内容只做删减不做修改。

本书融汇了最新的相关专业知识，图文并茂，浅显易懂；以工作任务为驱动，通过仿真化的学习情景来完成知识模块的学习，充分体现"实用为主，够用为度"的原则。本书由教学指导、学习材料、项目活动与任务、单词、词组和对话练习专业英语相关语法知识等组成。

本书为高等职业本专科院校机械大类专业英语教材，也可作为开放大学、成人教育、自学考试、中职学校、培训班的教材，以及企业工程技术人员的参考书。

本教材配有免费的电子教学课件和练习题参考答案，详见前言。

未经许可，不得以任何方式复制或抄袭本书之部分或全部内容。
版权所有，侵权必究。

图书在版编目（CIP）数据

机械制造类专业英语 / 杨成主编．—北京：电子工业出版社，2013.8（2023.2 重印）
高等职业院校精品教材系列
ISBN 978-7-121-20445-6

Ⅰ．①机… Ⅱ．①杨… Ⅲ．①机械制造－英语－高等职业教育－教材 Ⅳ．①H31

中国版本图书馆 CIP 数据核字（2013）第 103675 号

策划编辑：陈健德（E-mail：chenjd@phei.com.cn）
责任编辑：张　京　　文字编辑：张岩雨
印　　刷：北京虎彩文化传播有限公司
装　　订：北京虎彩文化传播有限公司
出版发行：电子工业出版社
　　　　　北京市海淀区万寿路 173 信箱　邮编 100036
开　　本：787×1 092　1/16　印张：13.25　字数：442 千字
版　　次：2013 年 8 月第 1 版
印　　次：2023 年 2 月第 12 次印刷
定　　价：42.00 元

凡所购买电子工业出版社图书有缺损问题，请向购买书店调换。若书店售缺，请与本社发行部联系，联系及邮购电话：(010) 88254888，88258888。
质量投诉请发邮件至 zlts@phei.com.cn，盗版侵权举报请发邮件至 dbqq@phei.com.cn。
本书咨询联系方式：chenjd@phei.com.cn。

前言

随着经济的全球化，中国已经成为"世界工厂"，对外交流的不断深入使社会对机械类外向型人才的需求日益迫切，因此《机械制造类专业英语》的出版是社会之所需。本教材根据行业技术发展新需求，按照目前最新的职业教育改革要求，结合示范专业课程建设经验和中澳职业教育合作项目成果，充分听取专业教师和行业专家意见进行编写。本书对传统的教学模式和课程设计进行了颠覆性的改革，具有一定的超前性，因此一定会发挥其应有的作用。

本教材注重英语通识能力和外向型人文思维能力的培养，如听、说、读、写、译等，尤其贯彻听说领先、读写跟进的原则，重点培养学生职场英语沟通与交流的能力，为机械行业输送用得上、下得去、留得住的实用型人才。本教材选取真实、典型的工作场景作为学习材料，充分落实以就业为导向、能力为本位、行业需求为目标，传授机械相关专业领域所需的职场英语知识和应用技能。本教材的教学采用教室（理论教学）+实作场（实践操作）的方式。

本书根据机械制造类专业教学的相应内容分为 4 大知识模块，并设计了大量的项目活动与任务。本书针对各个知识模块中的学习情景提供了大量的反馈练习，不同院校可根据实际情况对本教材内容做适当调整。

本书由重庆工业职业技术学院杨成编著，由机械工程学院副院长钟富平主审，机械工程学院的黄晓敏教授、姜秀华教授和赵平高级工程师给予了无私的指导，在此一并表示感谢。

本书的顺利出版，要感谢重庆工业职业技术学院中澳项目办、澳大利亚职业教育与培训鉴定服务中心（VETASESS）、职业教育与监督服务中心（VETASESS）；感谢 Allen Medley，Bruce Shearer 和 Vivien Carroll 等专家的大力支持；感谢电子工业出版社高职分社的领导给予的大力支持和帮助，并提出了许多宝贵的修改意见；本书曾参考并引用了有关文献资料、插图等，作者在此对相关作者表示由衷的感谢。

由于作者水平有限，时间仓促，写作中难免挂一漏万，不妥之处，恳请读者批评指正，并提出宝贵意见。

为了方便教师教学，本书还配有免费的电子教学课件、练习题参考答案，请有此需要的教师登录华信教育资源网（http://www.hxedu.com.cn）免费注册后再进行下载，若有问题请在网站留言或与电子工业出版社联系（E-mail：gaozhi@phei.com.cn）。

编　者

教 学 指 导

1. 教学方法建议

1）教学活动

活动名称 \ 知识模块	Module I～IV	活动操作程序	建议时间
小组讨论	√	1. 教师把设计好的问题（如是非判断，阅读理解等）用小纸条装在信封里，分发给各小组 2. 各小组讨论5分钟后派代表发言 3. 教师总结归纳	10～15 分钟
词汇游戏	√	根据图示识别单词、零部件	10 分钟
职场实地学习	√	实训场地，根据具体部件识别单词	20 分钟

2）Suggested Workplace Teaching Procedure

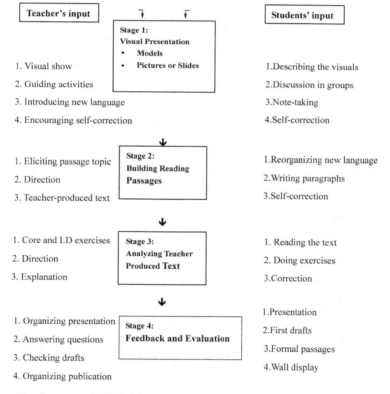

3）Teaching Equipments and Needs

（1）Multi-media equipments including projector, computer, screen, tape-recorder, microphones, etc.

（2）T&L materials (for teachers and students), teaching plan, lectures, teaching scheme, teaching reference books, etc.

（3）Teaching models, pictures, charts, course wares, etc.

2. Teaching and Learning Assessment Methods

modules \ methods	writing	spoken	observe	role-play	operation
Module I	√	√	√	√	
Module II~IV			√	√	√
words and expressions	√	√	√	√	

3. Teaching Assessments Methods

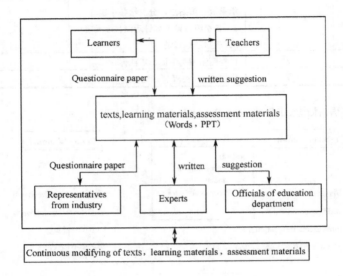

4. 教材中的图标介绍

scene	match words	required words
workplace	operation	true or false
compiled words	learning action	fill in blank
assessment	reading material	question
exercises	translation	group work
allocated time	exercises	

目 录

Module One　Basics in Mechanical Workplace ……………………………(1)
- Scene One　Health and Safety in Workshop …………………………………(1)
  - Grammar 1　句子成分 ……………………………………………………(11)
- Scene Two　Workplace Tools and Measuring Equipment …………………(12)
  - Grammar 2　科技英语词汇和句法的特点 ……………………………(34)
- Scene Three　Reading Skills of Scientific and Technical English …………(36)
- Scene Four　Product Specification …………………………………………(40)

Module Two　Mechanical Machining ………………………………………(44)
- Scene One　Mechanical Materials ……………………………………………(44)
  - Grammar 3　词类的转译 ………………………………………………(48)
- Scene Two　Electrical Machining ……………………………………………(50)
  - Grammar 4　翻译的基本方式及其选择 ………………………………(53)
- Scene Three　Lathe ……………………………………………………………(53)
  - Grammar 5　缩略词的解释 ……………………………………………(58)
- Scene Four　Milling ……………………………………………………………(60)
  - Grammar 6　短语的翻译 ………………………………………………(67)
- Scene Five　Radial Drilling Machines ………………………………………(68)
  - Grammar 7　分词 ………………………………………………………(72)
- Scene Six　Grinding Machines ………………………………………………(73)
  - Grammar 8　动词不定式 ………………………………………………(80)
- Scene Seven　Power in the Workshop ………………………………………(82)

Module Three　CNC Machining ……………………………………………(87)
- Scene One　CNC Machine ……………………………………………………(87)
  - Grammar 9　复合宾语 …………………………………………………(89)
- Scene Two　CNC Programming ……………………………………………(90)
  - Grammar 10　广告的翻译 ………………………………………………(95)
- Scene Three　CNC Machining Center ………………………………………(96)
  - Grammar 11　动名词 ……………………………………………………(102)
- Scene Four　CAD/CAM Software ……………………………………………(103)
  - Grammar 12　否定的处理 ………………………………………………(104)
- Scene Five　CAD/CAM Components and Functions ………………………(106)
  - Grammar 13　长句的翻译 ………………………………………………(111)

· VII ·

Scene Six　　FANUC Manual Guide ……………………………………………（114）
Module Four　　Die & Mold Tech ……………………………………………（124）
Scene One　　Transfer Molding ……………………………………………（124）
　　Grammar 14　数量词的翻译 ……………………………………………（126）
Scene Two　　Piercing and punching die ……………………………………（128）
　　Grammar 15　定语从句 ……………………………………………………（130）
Scene Three　　Injection Molding ……………………………………………（132）
　　Grammar 16　派生词的翻译 ……………………………………………（134）
Scene Four　　Bending Die ……………………………………………………（135）
　　Grammar 17　状语从句 ……………………………………………………（139）
Scene Five　　Extrusion ………………………………………………………（141）
　　Grammar 18　翻译的两个过程 …………………………………………（143）
Scene Six　　Drawing Die ……………………………………………………（144）
　　Grammar 19　and 引导的句型 …………………………………………（147）
Scene Seven　　Compression Molding ………………………………………（148）
　　Grammar 20　名词从句 ……………………………………………………（150）
Scene Eight　　Compound Die ………………………………………………（151）
　　Grammar 21　专业名词的翻译 …………………………………………（153）
Scene Nine　　Thermoplastic & Thermosetting Mold Design ………………（155）
　　Grammar 22　词义辨识法 ………………………………………………（156）
New Words ……………………………………………………………………（160）
References ……………………………………………………………………（204）

Module One

Basics in Mechanical Workshop

Scene One Health and Safety in Workshop

 Identifying hazards

What is a hazard?

A hazard is any situation with the potential to cause injury or illness. For example, situations that could pose hazards include: a system of work, a piece of machinery, a chemical that is used. To assist you in thinking about hazards, the following materials will break hazards into five major groups.

- Physical (e.g. plant/machine, noise, electrical, lighting, radiation, working at heights, housekeeping)
- Chemical (e.g. hazardous substances, dangerous goods)
- Ergonomic (e.g. manual handling)
- Psychological (e.g. work stressors)
- Biological

 Make a survey about hazards in your workplace.

Sample hazard inspection checklist

   Date of inspection: _____
   Area to be inspected: _____
   Checked by: _____

Floors

Safety issue	OK	Not OK	Plan to improve?
Even surface – no holes.			
Dropped objects picked up.			
Stock material out of way.			

Aisles

Safety issue	OK	Not OK	Plan to improve?
Wide enough for goods traffic.			
Adequately lit.			
Surface free from defects.			
Clear of cases, materials and rubbish.			

Fire

Safety issue	OK	Not OK	Plan to improve?
Extinguishers in place, recently serviced and clearly marked for type of fire.			
Adequate directions to fire exits.			
Exit doors easily opened from inside.			
Exits clear of obstructions.			
Fire alarms functioning correctly.			

General lighting

Safety issue	OK	Not OK	Plan to improve?
Adequate illumination.			
Good natural lighting.			
Good light reflection from walls and ceilings.			

Windows

Safety issue	OK	Not OK	Plan to improve?
Safety or reinforced glass where appropriate.			
Clean, admitting plenty of daylight.			
No broken panes.			
Ledge free of dust, tins or rubbish.			

Storage

Safety issue	OK	Not OK	Plan to improve?
Storage designed to minimize lifting.			
Materials stored in racks and bins wherever possible.			
Shelves free of dust and rubbish.			
Stacks stable with good foundations.			
Floors around racks clear of rubbish.			
Non-skid gratings in good condition.			

Module One Basics in Mechanical Workplace

Electrical

Safety issue	OK	Not OK	Plan to improve?
Gear not in use properly stored.			
No broken plugs, sockets or switches.			
Portable power tools in good condition.			

First aid

Safety issue	OK	Not OK	Plan to improve?
Cabinets and contents clean and orderly.			
Stretchers in position.			
Emergency numbers displayed.			

Chemical hazards

Safety issue	OK	Not OK	Plan to improve?
Are metal parts, paint brushes, etc. cleaned with turpentine, petrol or other solvents in uncovered buckets?			
Do people ever clean their hands with solvents of any type?			
Are there any hazardous chemicals in use on site?			

Ergonomics lifting/carrying

Safety issue	OK	Not OK	Plan to improve?
Do people have to lift heavy or awkward loads?			
Does the lifting involve high or low lifts?			

 The hierarchy of control of hazards in workshop.

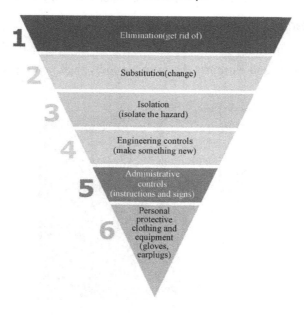

3

1. Elimination

Controlling risks through elimination may include, for example, eliminating toxic substances, hazardous machinery or processes that are not necessary to the way work is performed.

2. Substitution

Substitution may include, for example, substituting a toxic substance, hazardous machinery or process with one known to be less harmful to health or safety.

3. Isolation

Isolation may include, for example, enclosing or isolating a hazard such as a toxic substance by using a fume cupboard or using sound enclosure booths to control noisy machinery.

4. Engineering controls

Engineering controls may include changing processes, equipment or tools, for example:

- Installation of machine guards and machine operation controls.
- Ventilation to remove chemical fumes and dusts, and using wetting down techniques to minimize dust levels.
- Changing layout of work levels to minimize bending and twisting during manual handling.

5. Administrative controls

Administrative controls may include changing work procedures to reduce exposure to existing hazards. For example:

- Reducing exposure to hazards by job rotation.
- Limiting the number of people exposed to the hazard by limiting access to hazardous areas.

6. Personal protective clothing and equipment

Personal protective clothing and equipment include devices and clothing, such as safety clothing, footwear, helmets and earmuffs that provide individual employees with some protection from hazards.

Module One Basics in Mechanical Workplace

 Tick the correct response in the relevant box.　　　　True　False

(1) A hazard is any situation with the potential to cause injury or illness.　☐　☐

(2) Personal protective clothing and equipment are clothing and devices that provide employees with some protection from hazards.　☐　☐

(3) Administrative controls may include changing work procedures to reduce exposure to existing hazards.　☐　☐

 First aid

First aid is the provision of emergency treatment and life support for people suffering injury or illness.

First aid aims to:

- Promote a safe environment.
- Preserve life.
- Prevent injury or illness from becoming worse.
- Help promote recovery.
- Protect the unconscious.
- Reassure the ill or injured.

 Group discussion: contents list for a first aid kit.

Suggested contents for a first aid kit include:

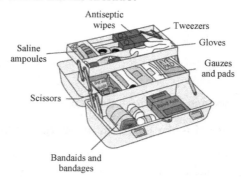

Sample contents list for a first aid kit

While the specific contents of a first aid kit will be determined from a risk assessment, the following contents list provides a basic outline.

- Emergency services telephone numbers and addresses, including local doctors and hospitals.
- Name and telephone number for workplace first aid.
- Basic first aid notes.
- Individually wrapped sterile adhesive dressings.
- Sterile eye pads.

- Sterile coverings for serious wounds.
- Triangular bandages.
- Safety pins.
- Small, medium and large sterile un-medicated wound dressings.
- Adhesive tape.
- Crepe bandages.
- Disposable gloves.
- Scissors.

 Group discussion: What should we have in a first aid room?

First aid rooms

Following are the types of contents that may be needed in a first aid room.
- Resuscitation mask.
- Sink and wash basin with hot and cold water supplied.
- Work bench or dressing trolley.
- Cupboards for storage of medical supplies, dressings and linen.
- Soiled dressing containers.
- Electrical power points.
- Medical examination couch with blankets and pillows.
- Upright chairs.
- Removable screens.
- Desk and a telephone.
- Stretchers.
- First aid kit.

Tick the correct response in the relevant box.   True   False

（1）The definition of first aid is the provision of emergency treatment and life support for people suffering injury or illness.   ☐   ☐
（2）One of the aims of first aid is to preserve life.   ☐   ☐
（3）First aid can be an effective alternative to medical treatment by a doctor.   ☐   ☐

 Suggested contents for a first aid kit include:

_____

Emergency management and fire safety

The types of emergencies that could occur at a workplace depend on factors such as the type of workplace and the location of the workplace.

Module One Basics in Mechanical Workplace

The types of emergencies that could occur include:
- Fire.
- Chemical spills or chemical release.
- Bomb threats.
- Structural faults in the building.
- Flooding or severe water damage.
- Threats of violence or robbery.

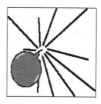

Examples of some of the emergencies

Fuel + heat/source of ignition + oxygen + chemical reaction = fire

Solvent + a spark + air + chemical reaction = fire

To prevent or extinguish a fire, you need to act on one of the four elements of fire:
- Source of ignition/heat.
- Fuel.
- Oxygen.
- Chemical reaction.

Examples of how to prevent fires include:

• Smoking only where permitted. Large, non-tip ashtrays should be used and everything in them should be cold before they are emptied.

• Keeping passageways and exits free from storage and waste in order to minimize combustible material.

• Promptly removing waste paper, packaging, old rags and other fire hazards.

• Ensuring that appliances (stoves, kettles. etc) are switched off each night. This should include computers and computer monitors if possible.

• Making sure that any broken electrical cords or plugs are replaced immediately. If an appliance or item of equipment smells or gives off smoke, turn it off, unplug it and do not use it again until it has been checked by a qualified technician.

机械制造类专业英语

● Ensuring there is plenty of air circulation space around heat-producing equipment (such as photocopiers and computers).

⚠ Tick the correct response in the relevant box.   True   False

（1）Only about 6% of oxygen needs to be present in the air for a fire to burn. ☐   ☐

（2）If an appliance or item of equipment smells or gives off smoke, turn it off, unplug it and do not use it.   ☐   ☐

（3）Smoking only where permitted.   ☐   ☐

（4）Ensuring that appliances are switched off each night. This may not include computers and computer monitors if possible.   ☐   ☐

👉 Workshop dangerous goods and strategies to control

Symbols for the different classes of dangerous goods.

What do these symbols mean?

Module One Basics in Mechanical Workplace

Dangerous goods are substances that may be corrosive, flammable, explosive, spontaneously combustible, toxic, oxidizing or water-reactive.

Effects on health and safety

The health and safety effects of dangerous goods are often immediate and severe.

Strategies to control dangerous goods

The preferred option to control risks arising from the use of dangerous goods is elimination.

Personal Protective Equipment (PPE) that includes overalls, aprons, gloves, dust masks, respirators, safety footwear, and goggles or face shields.

⚠ Tick the correct response in the relevant box.   True   False

(1) The preferred method of controlling risks arising from the use of dangerous goods is isolation.   ☐   ☐

(2) PPE stands for personal protective equipment.   ☐   ☐

 Look at this picture, what forms of personal protective equipment does the man use?

Tick the correct response in the relevant box.
☐ gloves ☐ overalls ☐ face shields ☐ goggles ☐ respirators

✍ Noise

A sound level meter being used to test noise levels.

Strategies to control Noise hazards

Personal Protective Equipment.

Examples of personal protective equipment for noise.

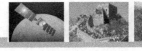

The following are some specific examples of ways to control noise:
(1) Eliminate or replace the plant or its operation with a quieter operation.
(2) Replace noisy equipment with new equipment that creates less noise.
(3) Use sound absorbing material to absorb vibration and thus reduce noise.
(4) Enclose noisy machinery or put barriers between the machinery and people.
(5) Reduce the amount of time people are exposed to noise, or use earmuffs.

Tick the correct response in the relevant box.   True   False
(1) A sound level meter can be used to test noise levels.   ☐   ☐
(2) Noise can only cause immediate hearing loss.   ☐   ☐
(3) Use sound absorbing material to absorb vibration and thus reduce noise. ☐   ☐

 Practice spoken English.

Li Ying: Good afternoon, Mr.Wang.
Mr. Wang: Good afternoon, Li Ying. Where are you going?
Li Ying: I am going to the library to borrow some books.
Mr. Wang: What kinds of books do you want to borrow?
Li Ying: I would like to borrow some books about health and safety in workshop.
Mr. Wang: Why?
Li Ying: I have been learning the course of machinery, but I have some trouble in health and safety in workshop. So I need some reference books.
Mr. Wang: What are the problems? May I help you?
Li Ying: Yes, what should we pay attention to in workshop?
Mr. Wang: Oh, well, first, we should identify hazards in workplace, and then ...The hierarchy of controlling emergency management and fire safety, next the Workshop Dangerous Goods and Strategies to Control, lastly the noise...
Li Ying: It seems very useful knowing…
Mr. Wang: Yes, you are right.
Li Ying: You are very kind, bye-bye.
Mr. Wang: Bye.

Module One Basics in Mechanical Workplace

New words and expressions we learn (let the students copy the new words and expressions he or she doesn't know in this scene and then dictate them among his/her group members).

Grammar 1 句子成分

Members of a Sentence
The members of a sentence mainly include subject, predicate, object, attribute and adverbial.

1．主语（subject）

主语是句子的主体，是句子所要说明的人或事物，主语通常是一些代表事物性或实体性的词语。如：

Bookkeeping is an essential accounting tool.
簿记是会计的基本工具。

除了名词可担任主语外，代词、数词、动词不定式、动名词、短语、从句均可作为主语使用，如：

She works at a big corporation.
她在一家大公司工作。

It is more convenient for us **to analyze the transactions**.
分析经济业务对我们来说是比较方便的。

It does not matter **what particular value is assigned to each share**.
每股的特定价值是无关紧要的。

2．定语（attribute）

定语是用来修饰名词或代词的。

A **small** business may employ only one bookkeeper.
小企业可以只雇用一名簿记员。

除形容词外，数词、名词所有格、动词不定式、介词短语、分词短语、动名词、副词从句都可作定语。

There is some **exciting** news in **today's** newspaper.
今天报纸上有令人振奋的消息。

Every Saturday Mr. Black goes to the supermarket to do shopping.
每星期六，布莱克先生都去超级市场购物。

3．谓语（predicate）

谓语说明主语的动作或状态，它的位置一般在主语后面。谓语是以动词为中心的，通常以动词词组表示。

All transactions **affect** at least two accounts.

所有经济业务至少影响两个账户。

Analyzing is the first step in the accounting process.

分析是会计程序的第一步。

4．宾语（object）

宾语表示动作的对象，是主语动作的承受者，有宾语的动词称为及物动词，宾语一般位于及物动词之后。作宾语的词有名词、代词宾格、数词、动词不定式、动名词、短语、从句等。

Mays purchased **office supplies** from the Clark Peeper Co.on account.

梅斯向克拉克皮帕尔公司赊购办公用品。

The balance sheet shows **the firm's condition** on the last day of the accounting period.

资产负债表表现会计期末企业的财务状况。

The debit and credit values of all transaction recorded in journal must be transferred to **the proper accounts** in the general ledger.

所有记录在日记账里的会计事项的借贷方金额，最终必须结转到总分类账的有关账户。

5．状语（adverbial）

状语是用来修饰动词、形容词或副词的，表示时间、地点、原因、方式、程度等。

Accounting is one of the fastest growing fields **in modern business world**.

会计是现代商业领域中发展最快的行业之一。

To explain the difference briefly, the accountant sets up a bookkeeping system.

为了简洁地解释差异现象，会计师建立起簿记系统。

If the bank lends him money, he must pay interest for its use.

如果银行贷款给他，他必须付利息。

Scene Two Workplace Tools and Measuring Equipment

Workplace Tools

Hand tools——General

pry bar 撬棍

Application: *Parting component, removing seals and aligning holes.*

brush 刷子

Application: *Washing components with solvent, removing dirt from parts.*

Module One Basics in Mechanical Workplace

wire brush　钢丝刷

Application: *Removing carbon rust gasket material dirt from parts.*

chisel　凿子

Application: *Cutting rivets bolts nuts and sheet steel.*

diamond-point cutter　钻石刻刀

Application: *Cleaning keyways threads and corners.*

die　板牙

Application: *Cutting external threads on rod or bolts.*

die wrench　板牙扳手

Application: *Holding and running a die during threading procedure.*

electric drill　电钻

Application: *Gripping and turning a twist drill.*

hand drill　手摇钻

机械制造类专业英语

Application: *As above.*

twist drill   钻头

Application: *Boring holes through metal, wood and plastic.*

twist conical drill   麻花锥形钻头

Application: *Removing broken studs(stud must be drilled through its centre).*

twist fluted and straight drill   直槽钻头

Application: *As above.*

chuck   凸轮

Application: *Removing broken chips but protruding stud.*

flat file   平锉刀

Application: *Restoring surface to a flat and smooth.*

half-round file   半圆锉刀

Module One Basics in Mechanical Workplace

Application: *Enlarging holes and rough finishing concave surface.*

round (rat tail) file　圆锉刀

Application: *A file with a circular cross section, used to enlarge small holes or file the inside of holes.*

small rat tail file　鼠尾细锉刀

Application: *Restoring point surfaces to condition.*

square file　方锉刀

Application: *Forming and finishing elongated notes.*

triangle file　三角锉

Application: *Restoring damaged threads .Sharpening saw teeth.*

portable grinder　便携磨具

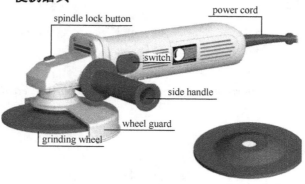

Application: *Reducing the size of metal parts. Cutting metal parts. Preparing metal for welding.*

机械制造类专业英语

angle grinder 角磨具

Application: *The angle grinder is a versatile cutting, grinding, polishing and sanding machine. It is traditionally used for cutting stone and metal, but has a multitude of other uses.*

hacksaw 手锯

Application: *Used for cutting most metal objects.*

round hammer 圆锤

Application: *Rounding off rivets and bolts. Forming metal. Driving drifts pins punches and chisels.*

hand sledge hammer 手用大锤

Application: *Driving large drifts pins, collars and shafts.*

soft hammer 软锤

Application: *Hammering parts made from materials that bruise easily or have a very hard surface.*

electric heater 电加热棒

Module One Basics in Mechanical Workplace

Application: *Joining steel or copper ware with resin cored solder.*

hexagon wrench 六角扳手

Application: *Loosening or tightening or set screws.*

lead light 照明灯

Application: *Illuminating the immediate work area.*

plier (combinations) 钳子

Application: *Twisting, gripping, bending and cutting most thin sheet metals or wires.*

long nose plier 尖嘴钳

Application: *Gripping and holding very small parts or fine wire.*

multi-diameter plier 多口径夹钳

Application: *Gripping and holding round parts.*

cutting plier 切割钳

Application: *Snipping wire, small bolts or rods and split pins. Removing and replacing split pins.*

ischaemum plier　鸭嘴钳

Application: *Removing and replacing internal and external chips.*

crocodile mouth plier　鳄嘴钳

Application: *Holding or clamping components.*

punching device　冲孔器

Application: *Aligning holes more parts.*

center punching device　中心冲孔器

Application: *Indenting metal surface to locate a twist drill.*

pin punch　针式冲头

Application: *Removing tapered or straight pins.*

starter   启动器

Application: *Breaking the initial grips of a pin. Used before a pin punch.*

scraper   刮刀

Application: *Removing carbon rust and stubborn dirt from flat surfaces.*

flat screwdriver   平口螺丝刀

Application: *Loosening and tightening screws with a standard slot in their heads.*

plum screwdriver   梅花螺丝刀

Application: *Loosening and tightening screws with a recessed cross in their heads.*

extension   加长杆

Application: *Locating the handle or speed brace in a position that clears other components or body work.*

handle-breaker bar　手闸杆

Application: *Initially loosening tight bolts or nuts.*

sliding wrench　滑动扳手

Application: *As above.*

ratchet　棘轮，棘齿

Application: *Unscrewing bolts or nuts when space around them is limited.*

universal joint　万向节

Application: *Allows a slight misalignment between a handle and socket.*

sleeve die (standard)　普通型套筒扳手

Application: *Used with a handle to loosen or tighten bolts or nuts.*

sleeve die (extra deep)　加深型套筒

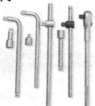

Application: *Loosening or tightening components or nuts fitted some distance from the end of their studs.*

Module One Basics in Mechanical Workplace

spark plug sleeve wrench　火花塞套筒扳手

Application: *Loosening or tightening spark plugs.*

impact sleeve　冲击套筒

Application: *Used with an impact wrench to loosen or tighten bolts or nuts.*

open end (flat)　双开口扳手

Application: *Loosening or tightening bolts or nuts located in a confined space.*

double plum wrench　双梅花扳手

Application: *Loosening or tightening bolts or nuts that have confined space above them.*

combination wrench　两用扳手

Application: *Has the advantages of both types of spanners.*

tap (hand)　丝锥

Application: *Using with a handle to cut a thread in a hole boring through or into most materials.*

tap wrench handle　丝锥扳手手柄

Application: *Holding and turning a tap during a threading process.*

T wrench　T 形扳手

Application: *As above.*

clamp　夹钳

Application: *Shearing or cutting most sheet metals or thin materials.*

toolbox　工具箱

Application: *Holding, separating and transporting hand tools.*

wrench (adjustable)　活动扳手

Application: *Substituting for the correct spanner.*

pipe clamp　管子钳

Application: *Gripping and turning cylindrical shapes.*
Hand tools—Specialized

grease gun 润滑油枪

Application: *Lubricating ball joints with grease.*

impact screwdriver 气动螺丝刀

Application: *Loosening extremely tight screws.*

impact wrench 气动扳手

Application: *Removing and replacing bolts or nuts at a fast rate.*

oil wrench 机油扳手

Application: *Removing an oil filter.*

oil gun (hand) 手动油枪

Application: *Withdrawing oil from a component or inserting oil into a component.*

pipe cutter 管子切割器

Application: *Cutting the end of steel or copper tubing so that it is square.*

pipe enlarge device (single & double) 管子扩孔器

Application: *Forming a uniform lip(flare) on the end of a steel or copper tube(pipe).*
Forming a double thickness lip(flare) on the end of a steel or copper tube(pipe).

piston ring compressor 活塞环压缩器

Application: *Clamping piston rings firmly into their piston grooves during the installation of the piston into its bore.*

rotary brush 旋转刷

Application: *Used with an electric drill to remove carbon from the combustion chambers and ports of a cylinder head.*

seal puller 油封拉拔器

Application: *Removing oil seals.*

steering wheel pulling device 转向盘拉拔器

Application: *Removing steering wheels.*

wheel cylinder clamp 车轮卡钳

Application: *Holding wheel cylinder pistons in their bores while the brake shoes and return springs have been removed.*

wheel (hub) puller 轮毂拉拔器

Application: *Removing hubs from spliced or tapered axle shaft ends.*

tie screw wrench 轮胎螺丝扳手

Application: *Loosening or tightening wheel nuts or screws. Those fitted with a blade are used to pry off hub caps.*

Hand Tools and Equipment

1. Type, Quality and Selection

An important part of the automotive repair industry is the type, quality and selection of hand tools and equipment. Hand tools are classed as those tools which can be held or supported by the operator's hands while they are being used. Equipment is defined as being those devices which are controlled by an operator but supported by the floor building or vehicle.

2. Service and Maintenance

When using all tools and equipment, the safety aspect is extremely important. Tools and equipment used for the purpose for which they were intended will not be a risk to the operator, his or her workmates or the component parts or the vehicle. It is the responsibility of the operator and the workshop manager to service and maintain all tools and equipment.

Three Groups: There are three groups related to the types of hand tools. Which include electrical automatic and mechanical? These groups are:

（1） General

- Used on a wide range of jibs;
- They are the responsibility of the trade people;
- Their length is limited to approximate 0.5m.

（2） Specialized

- Designed for a particular job;
- They are the responsibility of the workshop manager;
- Their length may be greater than 0.5m.

（3） Measuring instruments

- These are very accurate measuring device;
- They are responsibility of workshop manager. Although some trades people do supply their own;
- They are generally short than 0.5m.

Good quality hand tools and equipment will make sure that the jobs in the workshop can be completed without delay and to a high standard. Before an item is purchased, care must be taken that it will:

—do the required job (suitability);

—last for a long time (durability);

—operate correctly each time it is used (reliability).

The ability to quickly select the correct hand or equipment to complete a job successfully is achieved through practice. The aim of tool selection is to make ready all the tools needed for the job before work is started. This reduces the time spent "looking for tools".

Workplace Measuring Equipment

small air ball gauge　小空球量规

Application: *Used with an outside micrometer for measuring the inside diameters of small*

holes or bores.

water temperature detector　水温检测仪

Application: *Testing a cooling system for leaks. Testing a radiator cap pressure setting.*

pressure gauge　压力表

Application: *Testing the pressure created during the compression phase in an engine.*

dial indicator　百分表

Application: *Measuring small distances between* 0.01mm *and* 10mm, *for example a bend in a crankshaft (very accurate).*

external micrometer　外径千分尺

Application: *Measuring external diameters or lengths from* 0.01mm *to* 100mm, *for example the diameter of a piston skirt (very accurate).*

feeler gauge　塞尺

Application: *Measuring small clearances or gaps that exist between two components. For example used to measure tappet clearance.*

inside caliper　内卡钳

Application: *Used with a steel rule for measuring internal dimensions of components. For example, measuring the inside diameter of exhaust tubing.*

inside micrometer 内径千分尺

Application: *Used with an external micrometer for measuring the internal dimensions of components, for example to measure the diameter of a cylinder bore (very accurate).*

oil pressure gauge 油压计

Application: *Testing the oil pressure of running engine.*

outside caliper 外卡钳

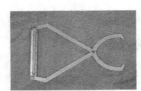

Application: *Used with a steel rule for measuring the external dimensions of components. For example to measure the outside diameter of exhaust tubing.*

pressure vacuum gauge 真空压力表

Application: *Testing the pressure and vacuum of a mechanical fuel pump. Testing the inlet manifold vacuum of a running engine.*

sparking-plug gauge 火花塞规

Application: *Measuring the gap that exists between the of a spark plug. Adjusting the gap of a spark plug.*

steel ruler 钢尺

Application: *Measuring the dimensions of component(least accurate of measuring devices).*

steel tape 钢卷尺

Application: *Measuring distances that are greater than one metre.*

straight-edge ruler 直尺

Application: *Used with a feeler gauge for measuring the twist or buckle(distortion) of a flat surface.*

electric tension wrench 电子扭矩扳手

Application: *Setting the tension of bolts or nuts to a given value.*

vee block V 形块

Application: *Used with a flat surface for supporting a round or cylindrical component during a measuring task. For example checking a camshaft for bend.*

vernier caliper　游标卡尺

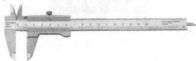

Application: *Measuring dimensions of most types of shapes and depth of holes to within 0.02mm(very accurate).*

electrical ammeter　电流表

Application: *Measuring current flow in a given electrical circuit.*

battery discharge tester　蓄电池放电检测仪

Application: *Testing the condition of a battery by placing it under load.*

hydrometer tester　液压表

Application: *Testing the specific gravity of the electrolyte in various cells of a battery.*

low voltage tester　低电压表

Application: *Checking the existence of a voltage supply to an electrical component.*

ohmmeter 电阻表

Application: *Testing the resistance of an electrical component that has been disconnected or removed from its circuit.*

multi-meter 万用表

Application: *Testing amps, ohms and volts related to the components in a given electrical circuit.*

voltmeter 电压表

Application: *Measuring the voltage drop across an electrical component when the circuit is turned on.*

air compressor 空气压缩机

Application: *Pressuring and storing air to ensure a steady supply at a 550~850kps.*

battery charger 充电器

Application: *Charging a battery either in or out. Charging several batteries out of the vehicles.*

bench grinder 台式砂轮机

Application: *Removing material from small metal components. Sharpening hand tools.*

electric welder 电焊机

Application: *Welding most metals with an electrode held in its hand piece.*

All service and repair jobs require analysis, diagnosis and measurement to determine the cause of the trouble. After disassembly, a regular series of steps are repeated to determine each component's serviceability. To reassemble and install engine components, make exact measurements to ensure proper operation.

1. Types and Uses

In the S.I. metric system of measurement, the unit of length is the metre. In automotive work the most commonly used unit of length is the millimeter, which is one thousandth (1/1000) of a meter. For more precise and smaller measurements the millimeter is divided into tenths and hundredths.

Because accurate measurement is often essential, it is necessary to learn to use the various measuring tools efficiently.

The simplest tool for measuring is the graduated rule which, for automotive work, is made of alloy steel, hardened and ground.

2. Care and Storage

A measuring instrument is a delicate precision instrument and must be treated accordingly. To maintain consistent results and the accuracy of the instruments, the following points should be observed.

Module One Basics in Mechanical Workplace

—before using the instrument, clean the faces.

—check the zero setting before starting the measurement.

—switch off electrical meters when not in use and keep them away from heat and moisture.

—clean and steady the work piece or component being measured.

(1) When they are not in use, place them back in their box or plastic wallet.

(2) Store them in a safe place. For example, in a cupboard drawer, on a shelf or in a safe area on the workbench.

(3) Before putting them away, clean the instruments and where necessary lightly oil the moving parts.

(4) Storing instruments in such a manner as will ensure that they are safe from dust, corrosion and accidental damage.

3. Types and Uses—Electrical

Test instruments

Many of the component operations on a vehicle are physical. The method of operation is visible and can be readily understood. When considering electrical and electronic systems, the problem becomes more difficult. The effect of electrical current is observable—a starter motor turns or a lamp is illuminated—but electricity cannot be 'seen' in the same way as it is possible to observe the flow of oil or fuel.

In order to test electrical circuits and components, a means of measuring the flow rate (current), the flow pressure (voltage) and the resistance to flow (resistance) of electricity is required. The majority of electrical system testing can be achieved by the use of a voltmeter, an ammeter and an ohmmeter. The instruments can be analogue (moving pointer on a printed scale) or digital (numbers appearing on a liquid crystal display). Selection of a test instrument should be based on the type of work it will be used to test. Relatively cheap moving coil meters are adequate when high current flow is present and / or voltage reading accuracy is not critical. Extremely accurate moving coil meters are available, but these tend to be expensive. The type of meter gaining popularity amongst mechanics and technicians is the digital multi-meter. These units are robust and many are auto-ranging (eg. not necessary to change scales for accuracy). They are overload protected in some ranges and can be switched to read amps, volts and ohms. The accuracy of these units on a cost basis is superior to analogue meters.

The three forms of electrical circuits are:

(1) Series.

(2) Parallel.

(3) A combination of both, known as series/ parallel.

A working knowledge of these circuits, coupled with an understanding of the definition of the volt, ampere and the ohm will be of invaluable assistance when using instruments for testing.

Practice spoken English.

Miss Li: Hi, Mr. Liu.

Mr. Liu: Hi, Miss Li, how are you?

Miss Li: I am fine. Thank you. But I have some problems with the workplace tools and measuring equipments. Could you give me a favor?

Mr. Liu: Yes, my pleasure. What are these problems?

Miss Li: Can you tell me some of the workplace tools and measuring equipments?

Mr. Liu: Oh, there are many kinds of workplace tools and measuring equipments in common use, such as seal puller, steering wheel, valve tapper, valve seat cutter, valve seat hone…

Miss Li: What are they used for?

Mr. Liu: The seal puller are used for removing oil seals and valve seat cutter are used for restoring valve seats in the combustion chambers of a cylinder head.

Miss Li: Are the…used in mass production?

Mr. Liu: No.

Miss Li: Why?

Mr. Liu: Because they are rather…in operation.

Miss Li: I am really appreciated you for telling me these.

Mr. Liu: I was glad to be of some services.

New words and expressions we learn (let the students copy the new words and expressions he or she doesn't know in this scene and then dictate them among his/her group members).

Grammar 2　科技英语词汇和句法的特点

1．词汇的特点

（1）常用词汇专业化。

Marketing：市场→销售

（2）同一词语词义多专业化。

book：书→账本，订购，售票

（3）广泛运用构词法，形成丰富多彩的专业词汇。

A．合成法（Compounding）

所谓合成法，就是将两个或两个以上的旧词合成一个新词。

B．混成法（Blending）

所谓混成法，就是选取两个词中在拼写上或读音上比较合适的部分，而其余部分除去。通常是前一词去尾，后一词去首，然后叠加混合。

C．词缀法（Acronym）

所谓词缀法，就是取出某一词语中所有主要词的第一个字母组成新词，缩略法组词最典型的例子，就是大家熟悉的词 AIDS，它是由 acquir immune deficiency sydrome 这个词语组合缩略而成的。

2．词法的特点

科技英语在词法方面的显著特点是其名次化倾向。

（1）广泛使用表示动作或状态的抽象名词和动名词。

（2）广泛采用名词连用形式

所谓名词连用，是指中心名词前有许多不变形的名词，充当其前置形容词修饰语。英语语法中称其为"扩展的名词前置修饰语"。如：

Customer service 客户服务

（3）普遍采用以名词为中心构成的词组表达动词概念。如：

go fishing 钓鱼

采用以名词为中心词词组表达动词概念。可使文中的谓语动词形式多样，增加行文的动感。

3．句法的特点

1）广泛使用被动语态

与汉语相遇，英语的被动语态本来就用得很普遍，再加上科技英语叙述的对象往往是事物、现象或过程，注重的是其叙述的客观事实，强调的是其所叙述的事物本身，而并不需要过多注意它的行为主题是什么。这时，使用被动语态不仅比较客观，而且可使读者的注意力集中在所叙述的客体上，而这正是科技英语所需要的。

此外，被动语态句的使用，还可避免给人以主观臆断的印象，而这也是科技英语所需要的。

2）长句使用过多

相对于汉语而言，英语本身长句就比较多。而在科技英语中。又常常需要表达多重密切相关的概念。同时，科技英语文体还特别讲究推理严谨和叙述准确。因此，句中修饰成分、限制成分甚多，这就必然形成长句。例如：

Intensive distribution is appropriate for products such as chewing gum, candy bar, soft drinks, bread, film and cigarettes where the primary factor influencing the purchase decision is convenience.

集约型配送适用于口香糖、糖果、软饮料、面包、胶卷和香烟等影响购买决定的主要因素是其方便性的产品。

3）广泛使用包括非谓语动词短语在内的各种短语

如上所述，科技英语中的句子较长，子句较多。但若子句过多，整个句子就会显得冗长。为使句子结构简洁，避免或减少复杂的长句，故多采用各种类型的短语来替代。

4．修辞的特点

1）时态运用有限

科技英语所运用的时态大都限于一般现在时、一般过去时、现在完成时和一般将来时这几种，其他时态很少运用。

2）修辞手法单调

与汉语一样，英语也有很多修辞手法。通常有夸张（Hyperbole）、明喻（Simile）、隐喻（Metaphor）、借喻（Metonymy）、拟人（Personification）和对比（Contrast）等。这些手法在英语的文学体中是常见的，但在科技英语中却少见。

3）普遍使用逻辑性语法词（Logical Grammatical Operators）

Scene Three　Reading Skills of Scientific and Technical English

科技英语文章信息含量大，大部分内容是陈述事实，读这类文章时不仅要提取信息，还要充分理解材料。提高科技英语文章的阅读能力有以下几个方面的技巧。

1．获取主题思想（Reading for Main Idea）

我们可以把获取主题思想的阅读技巧分四步。

（1）辨认主题名词；
（2）找出主题句；
（3）获取主题思想；
（4）避免不相关的内容。

2．获取细节（Reading for Details）

在所有文章里，作者都使用细节或事实来表达和支持他们的观点。阅读要想有效果，就要能辨认并记住文章中重要的细节。一个细节就是一个段落中的一条信息或一个事实。它们或者给段落的主题提供证据，或者为其提供例子。有些细节或事实是完整的句子，而有些只是简单的短语。怎样确定什么是细节呢？你可以问自己这样的问题，这是我能讲给别人听的细节吗？例如，"he was ill"和"Dogs have fleas"这些都是细节；而"of the city"和"putting on a hat"就不是细节，因为它们不包含你能够讲给别人听的信息。

另外只是判断出哪些是细节往往并不够，在很多情况下，还必须能区分哪些是重要细节，哪些是次要细节，在阅读时要尽量发现并记住它们。

3．理解词义（Understanding Vocabulary）

虽然阅读时不必认识每一个生词，但词汇量的多少会影响一个人对文章的理解力，而且科技英语词汇和普通英语词汇有很大的不同，学生平时很少接触，看起来难度较大不好理解。所以扩大词汇量是提高阅读能力的一条切实可行的重要途径。另外如何正确选择单词的意思，在很大程度上也影响对于文章的理解，对此有三个技巧。

1）通过上下文线索猜测词义

在科技英语文献中，作者为了阐述本领域的一个概念、实验过程，会用到一些专业术语，为了使读者更好地理解作者作品的内容和实质，作者对那些读者不熟悉的或有不同解释的词汇、术语就会用下定义的方法。定义在科技英语文献中使用非常频繁，那么识别定义，对猜测词义和概念大有裨益。定义可由破折号、逗号、圆括号等标点符号引出，也可由that is, meaning, such as, namely,　in other words, that is to say, for example, known as 等词语引出，例如：

If a tooth on a helical gear makes a complete revolution on the pitch cylinder, the resulting gear is known as a worm.（如果螺旋齿轮上的一个齿在节距圆柱上绕一整圈，这样形成的齿轮就叫蜗杆。）

句中用了 known as 作为对 worm（蜗杆）的定义信号。

除了使用定义猜测词义外，还可根据读者的经验和常识猜测词义。一个人的阅历越丰富，知识面越广，猜测生词的能力就越强，例如：

Mark got on the motorbike, and I sat behind him on the pillion and we roared off into the night.

根据我们的日常生活经验，可以猜出 pillion 的意思是"摩托车后座"。

另一种猜测词义的方法是比较和对照的方法。作者有时为了帮助读者理解文章，将生词与熟悉的词进行比较。例如：

Doctors believe that smoking cigarettes is derimental to your health, They also regard drinking as harmful.

第二中的 They，also，drinking 和 harmful 与第一句中的 derimental 发生联系，帮助我们猜出 derimental 的词义是"有害的"。

2）通过构词法掌握词汇前缀、后缀和词根

如果较好的掌握了构词的三要素：前缀、词根和后缀的话，你的词汇运用能力就会得到很大提高，科技英语的读者更会从中受益匪浅，因为许多科技英语词汇都是从希腊语和拉丁语的词根派生出来的，而且，相当数量的科技领域里的新词和术语，都是遵循构词法的。

通过分析构词法，可以了解每一部份的意思，进而理解整个词的意思。例如：我们来猜测"contradict"的词义。

contradict = contra + dict

contra = meaning "against, opposite", dict = meaning "say, speak"

因此可以推测这个词的意思应该就是"speak against"，或者是"say the opposite"

下面列出一些最常用的前缀。

a-	(without)	amoral
ab-	(not, away from)	abnormal
anti-	(against)	antibodies
bi-	(two)	bilingual
bio-	(life)	biochemistry, biology
circum-	(around)	circumference
co-	(together)	coefficient
counter-	(against)	counteract
contra-	(against)	contradict
dis-	(not)	discover
ex-	(former)	ex-chairman
ex-	(out of)	exposed, exterior
extra-	(outside)	extraordinary
fore-	(before)	foresee

续表

il-	(not)	illogical
im-	(not)	impossible
in-	(not)	inadequate
inter-	(between)	interchangeable
intra-	(within)	intramuscular
ir-	(not)	irregular
mal-	(bad)	malnutrition
mega-	(big)	megahertz
micro-	(small)	microscope
mid-	(middle)	midair
mini-	(small)	minicar
mis-	(not)	miscalculate
multi-	(many)	multinational
non-	(not)	nonreader, nonsurgical
over-	(more than necessary)	over, overlearn
post-	(after)	postpone
pre-	(before)	precede
re-	(again)	replace
sub-	(under)	substandard
trans-	(across)	transplant
un-	(not)	unhealthy

3）扩大知识面，克服背景知识方面的障碍

除词汇外，科技英语阅读的另外一个困难因素是背景知识。阅读理解有两种图式：语言图式和知识图式。知识图式（schemata）是指用背景知识去理解阅读材料所传递的信息，也就是用读者头脑里的固有知识去理解、消化、吸收材料的内容。读者在阅读英语书籍时，往往由于对文章所涉及的历史、文化、政治、经济等方面的知识缺乏了解而读不懂，或者是由于材料所涉及的专业知识超过了其指示范畴而导致理解困难。

4．理解句子（Understanding Sentence）

阅读一篇文章时，经常会遇到这样的情况：没有生词但却读不懂整句话的意思，特别是遇到长句或较为复杂的句子时，这时可以利用下面四个技巧：

（1）分析句子。当读到不懂的句子，可以考虑把句子拆开，分别找出主语、谓语动词和宾语，最后再看修饰语。

（2）利用标点符号。和词一样，标点符号同样可以传达作者的意思，因此有必要了解标点符号的意思和用法，并培养自己运用标点符号确定单词和句子意思的能力。

（3）利用参照词。科技文章常用参照词，这样可以避免一个词的重复使用，这些参照词包括人称代词如：it、they, he 等，指示代词如：which、that、who 等，还包括名词等。如果读者不

 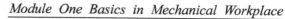

能识别这些参照词，就会妨碍阅读理解，所以要学习确定每一个参照词在文章中指代的是什么。

（4）利用信号词。信号词可以连接词与词，也可以连接句子与句子，或者段落与段落。如果不理解这些信号词，就会影响阅读理解，因此，要特别注意这些词并尽力理解它们所代表和指代的意思。常用的信号词有下列几种。

表示附加：and, as well as, besides, apart from, in addition to, furthermore, moreover 等。

表示因果：owing to, accordingly, due to, hence, in view of, as a result, thus, therefore, so 等。

表示对照：but, though, although, yet, in contrast 等。

表示条件：if unless, provided 等。

表示强调：above all, particularly, especially, in particular 等。

5. 抓住段落核心，整体理解文章（Understanding Article）

科技英语阅读的目的是为了理解文献的内容并获取所需要的信息。在阅读时，同样可以采用一般英语阅读中所使用的研读（study reading）、略读（skimming）、浏览（scanning）等阅读技巧。

1）研读

科技文章的内容是浓缩的，你会在一页上发现大量的信息，有不少学生都会有这种感觉，在读完一篇英语科技文章后，总是停留在文字的表面意义上，弄不清作者究竟讲了些什么。

阅读英语科技文章和阅读小说、报纸等不一样，一定要放慢速度，同时记笔记或记卡片，而且要经常复习所读过的段落。

2）略读

在时间有限的情况下，就要求对文章内容有个总体的了解，这时常常需要进行略读。与浏览不同，略读不需要寻找特定的数目和名称，只是制定主题，所以，进行略读的一种方法就是判定可能的主题句。

3）浏览

浏览可以帮助你得到你想得到的特定信息，在你已经知道一篇文章的大概内容后，而你又想从中得到你对特定问题的答案时，就可以用浏览的方式。

Exercise1

Determine whether the following underlined words are details.（Yes or No）

a）The climate of this area is typical of desert conditions. Summer temperatures often do not vary widely from day to night. (   )

b）Ferns love moisture and shade. If at all possible avoid putting them in full sun. (   )

c）The best thing to do the night before a big test is get a good night's rest. (   )

d）Farm was the main occupation of the people of this area ; using irrigation they grew corn, beans, and squash. (   )

Exercise2

Choose the correct lettered response to complete each numbered statement.

1）In inadequate, inanimate, infrequently, in-means _____.

    a. interior   b. not

2) In preceded and predict, pre- means _____.
   a. for   b. before
3) In replace, renew, and reproduce, re- means _____.
   a. again   b. strong
4) In malnutrition, mal- means _____.
   a. bad   b. animal
5) In antibiotic, anti- means _____.
   a. before   b. against
6) In biology and biochemistry, bio- means _____.
   a. life   b. two
7) A subnormal temperature is _____.
   a. below normal   b. above normal
8) To circumnavigate an object means _____.
   a. to move around it   b. to move across it
9) In minicar, mini- means _____.
   a. small   b. big
10) Microsurgery involves _____.
    a. a small area of the body   b. the whole body

Scene Four   Product Specification

产品说明书是帮助用户认识产品，指导用户使用产品的书面材料，它对产品的结构、功能、特性、使用方法、保养、维修，注意事项等做出详细的解释说明，既介绍产品又传授知识和技能。

由于产品的种类、性质不同，其说明方法、内容也不同，机械产品说明书的内容一般包括产品特点、用途、规格、结构、性能、操作程序及注意事项等。

由于说明书文体旨在介绍使用方法、操作方式或注意事项等，因此它的文体特点为简略性、技术性、描述性和客观性。简略性表现为内容简略，文字浅显，句式简单，大量使用祈使句和省略句，避免使用不必要的修辞手段。说明书中文字大多涉及某方面的专业知识，即使是一般家用电器的使用说明书也具有技术性，工用机械说明书技术性就更强。描述性表现为文字具有层次感、程序性和说服力。产品说明书主要是让消费者了解产品的性能、特点等，因此语言客观，用在说明书中的词语应该恰当、如实地反映产品的真实情况。

产品说明书的式样多种多样，但在写法上有许多共同的地方。一般说来，产品说明书大多都由标题、正文和落款三部分组成。

产品说明书不同于其他文体，我们需要掌握其文体和语言等特征，用符合说明书的语言如实地将产品说明书的内容翻译出来，其次还需要了解产品的基本原理等相关专业知识。请参阅下例。

Module One Basics in Mechanical Workplace

Introduction of Philips Desk Light

Thank you for your purchase of a Philips desk light!

Caution Warnings

For you protection, please read this manual and the safety warnings carefully before using the desk light and keep it for later reference. The manufacturer will not be held responsible for any damages caused by improper usage or modifications to the desk light.

Electrical Specifications

Voltage/Frequency:	220V~50Hz
Power/Lamp:	27W/PL-F 27W/4P
Lampholder:	GX 10q-4
Electrical Insulation Classification:	II

Operation Instructions

This desk light is not waterproof and is only suitable for indoor usage.

No alterations of any kind should be carried out on this desk light.

Do not use any voltage exceeding a 10% margin of the specified standard.

Do not touch the lamp inside or the lamp shade when the power is on.

Do not place flammable material near the desk light.

In case of operation failure , please switch off the desk light , unplug the power cord and contact your nearest Philips dealer.

This desk light is suitable for Philips PL-F/4P 27W compact fluorescent lamps only.

Installation

Take out the desk light body, base and clip.

To use the desk light with the base, insert the stem onto the base plate and tighten the fixing screw.

To use the desk light with the clip, mount the clip onto the desired surface and tighten the fixing screw , then insert the stem onto the clip.

The supplied clip is not suitable for use on tubes.

Maintenance

In order to ensure optimum performance , we recommend you to clean the desk light twice a year. When cleaning the desk light take care to use soft cotton cloths only.

Do not use chemical solvents to clean the desk light, it might damage the painted surfaces.

Caution: Please turn off the power and unplug the desk light before replacing the lamp!

Complete the following sentences with the appropriate expressions given below. Change the form where necessary.

switch on	switch off	turn on	turn off	be suitable for

(1) As he entered the dark basement, he _____ the light.

(2) He _____ the road when he ought to have gone straight.

(3) He _____ the radio to listen to the news.

(4) You should _____ the light when you leave the room.

(5) This desk light is only _____ indoor usage.

 Translate the following sentences taken from the passage into Chinese.

(1) For you protection, please read this manual and the safety warnings carefully before using the desk light and keep it for later reference.

(2) The manufacturer will not be held responsible for any damages caused by improper usage or modifications to the desk light.

(3) In case of operation failure, please switch off the desk light, unplug the power cord and contact your nearest Philips dealer.

(4) To use the desk light with the clip, mount the clip onto the desired surface and tighten the fixing screw, then insert the stem onto the clip.

(5) In order to ensure optimum performance, we recommend you to clean the desk light twice a year. When cleaning the desk light take care to use soft cotton cloths only.

Reading the following introduction of shaver and trying to translate the passage into Chinese.

CHARGING

*Recharge and keep the appliance at a temperature between 5℃ and 35℃.

*The appliance is suitable for mains voltages ranging from 220V to 240V.

*To optimize the life of the appliance, do not leave the shaver connected to the mains continuously.

(1) Slide out the slide pin.

When charging, make sure the appliance is in OFF position.

(2) Put the appliance into the socket.

The green pilot light will light up.

Charging normally takes approx.8 hours.

(3) Remove the appliance from the socket when the battery has been fully charged. A fully charged shaver has a shaving time of up to 30 minutes.

(4) Retract the slide pin.

SHAVING

(1) Switch the appliance on by pressing the switch lock and push the ON/OFF button upwards.

The switch lock will prevent accidental switching-on of the appliance.

(2) Move the shaving heads quickly over the skin, making both straight circular movements.

Shaving on dry face gives the best results.

Your skin may need 2~3 weeks to get accustomed to the shaving system.

Module One Basics in Mechanical Workplace

(3) Switch the appliance off.

(4) Put the protective cap on the shaver to prevent damage.

Replace the shaving heads (type HQ 3 Double Action) every 2 years for optimal shaving results.

CLEANING

*Regular cleaning guarantees optimal shaving performance.

*Make sure the appliance is switched off and disconnected from the socket when you clean the appliance.

*Clean the shaving unit every week.

(1) Clean the top of the appliance with the supplied brush.

(2) Press the release button and remove the shaving unit.

(3) Brush the inside of the shaving unit and the hair chamber.

(4) Put the shaving unit back onto the appliance.

 Self assessments

Time_____ Class_____

No.	Name	Evaluation Items				Notes
		Attitude	Presentation& Passage	Exercises	Activities	

Teachers' Observation Record and Evaluation Sheet

level \ item	listening	speaking	reading	writing	familiarity of parts
good					
better					
best					
bad					

Module TWO
Mechanical Machining

Scene One Mechanical Materials

Before you step into this scene, please learn the following words and expressions by heart with your classmates by looking up the dictionary or surfing the internet.

Look at the following words we may use in this scene and match the English and Chinese meaning.

supervision	n. 氧化物
flexibility	n. 碳酸盐，黑金刚石
sulfide	v. 精炼，提纯
optimum	a. 无穷的 n. 无尽
cam	n. 硫化物
metallic	n. 机动性，灵活性
carbonate	n. （复数为 optima）最适合条件
refine	n. 车床
commercial	a. 有机的
oxide	a. 金属的
nonmetallic	n. 凸轮
lathe	a. 特有的 n. 特性
infinite	a. 非金属的
organic	n. 监督，管理
characteristic	a. 商业上的，商业的

Module Two Mechanical Machining

  The following expressions we may use, before we finish off the task please make out their Chinese meaning with the help of your teammates or dictionary or surfing the internet.

   special-purpose   differ from…
   general-purpose   either… or…
   nonferrous metal   not only … but also …
   suite to

The material and the process are understood in the design and manufacture of a product. <u>Materials differ widely in physical properties, machinability characteristics, methods of forming, and possible service life.</u>

 Mechanical materials fall into two basic types: metallic or non-metallic. Non-metallic materials are further classified as organic or inorganic substances. <u>We should pay more attention to choosing the appropriate material because there is an infinite number of non-metallic materials and pure as well as alloyed metals.</u>

 <u>Few commercial materials such as oxides, sulfides, or carbonates, exist as elements in nature.</u> So they must undergo a separating or refining operation before they can be further processed. In metal working, iron is the most important natural element. When combined with other elements into various alloys it becomes the leading engineering metal. <u>The nonferrous metals, including copper, tin, zinc, nickel, magnesium, aluminium, lead, and others all play an important part in our economy.</u>

Make a survey about mechanical engineering materials in your workplace.

Task One

List the factors that form the differences among the mechanical engineer materials after discussion with your group members.

Task Two

Please fill in the forms about the types of engineering materials.

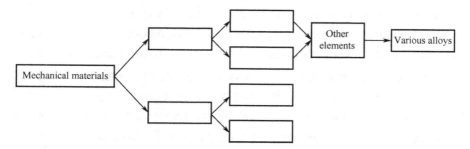

  Read the above underlined materials again with your workmates and then try to translate them into Chinese, and then discuss with your workmates again.

(1) Materials differ widely in physical properties, machinability characteristics, methods of forming, and possible service life.

(2) We should pay more attention to choosing the appropriate material because there is an infinite number of non-metallic materials and pure as well as alloyed metals.

(3) Few commercial materials such as oxides, sulfides, or carbonates, exist as elements in nature.

(4) The nonferrous metals, including copper, tin, zinc, nickel, magnesium, aluminium, lead, and others all play an important part in our economy.

After talking about the materials with your group members and then do your best to answer the following questions according to the above materials.

(1) What facts should the designer consider in selecting an economical material and a process that is best suited to the product?

(2) Which natural compounds of metals do not exist as elements in nature?

(3) Can you list some nonferrous metals according to the above materials?

Manufacturing requires tools and machines that can produce economically and accurately. Economy depends on the proper selection of the machine or process that will give a satisfactory finished product, its optimum operation, and maximum performance of labour and support facilities. In small-lot or job shop manufacturing, general-purpose machines such as the lathe, drill press, and milling machine may prove to be the best because they are adaptable, have lower initial cost, require less maintenance, and possess the flexibility to meet changing conditions.

A simple bolt may be produced on either a lathe or an automatic screw machine. The lathe operator must not only know how to make the bolt, but also be sufficiently skilled to operate the machine. On the automatic machine the sequence of operations and movements of tools are controlled by cams and stops, and each item produced is identical with the previous one. Often it is uneconomical to make a machine completely automatic. Because the cost may become prohibitive, most parts can be produced by several methods, but usually there is one way that is most economical.

Factors that must be considered are volume of production, quality of the finished product, and

Module Two Mechanical Machining

the advantages and limitations of the equipment capable of doing the work. Most parts can be produced by several methods, but usually there is one way that is most economical.

Read the information above with the help of your classmates, try to understand the meaning of it and then fill in the blanks in the following short paragraph.

We should pay attention to two elements such as _____ and _____, when considering tools and machines in manufacturing.

We must think of these factors if we have to manufacture economically, they are the _____ of a machine and the process which produce the products, its _____ operation, and _____ performance of labour and _____.

Discuss with your colleagues and try to share the ideas of your partners and then make sure which of the following expression is right or wrong. T for Right and F for wrong.

( ) General-purpose machines may prove to be the best in small-lot or job shop manufacturing.

( ) Lathe, drill press, and milling machine may meet changing conditions since they are adaptable, have lower initial cost, require less maintenance, and possess the flexibility too.

( ) To operate the machine, the lathe operator must know both how to make the bolt and be sufficiently skilled.

Survey your factory attached to your college and make it clear that why it is uneconomical to make a machine completely automatic, talking about it in your group.

_____

Read the last paragraph of the materials above and do your best to answer the following questions with the help of your classmates.

(1) When considering economic factors, what kind of elements should we pay more attention to?

_____

(2) Why lathe, drill press, and milling machine may prove to be the best that can produce economically and accurately?

_____

Read the underlined materials above again with your workmates and then try to translate them into Chinese, and then discuss with your workmates again.

(1) On the automatic machine the sequence of operations and movements of tools are controlled by cams and stops, and each item produced is identical with the previous one.

_____

(2) Most parts can be produced by several methods, but usually there is one way that is most

economical.

Practice spoken English.

Mr. Liu: Good afternoon, Mr. Xie.
Mr. Xie: Good afternoon, Mr. Liu
Mr. Liu: May I ask you a few questions?
Mr. Xie: Yes, of course. What are the problems?
Mr. Liu: How many kinds of mechanical materials are there in common use?
Mr. Xie: There are many kinds of mechanical materials.
Mr. Liu: What are they?
Mr. Xie: They are …
Mr. Liu: Are mechanical materials important in nearly all of machinery?
Mr. Xie: Yes, they are.
Mr. Liu: What can the mechanical materials be used for?
Mr. Xie: …
Mr. Liu: How do you distinguish the mechanical materials?
Mr. Xie: …
Mr. Liu: Thank you for telling me about these.
Mr. Xie: Not at all.

New words and expressions we learn（let the students copy the new words and expressions he or she doesn't know in this scene and then dictate them among his/her group members）.

Grammar 3　词类的转译

英语的一个词类能充当的句子成分较少，充当不同成分时需要改变词类；而汉语的一个词类能充当的句子成分较多，充当不同句子成分时无需改变词类。因此英汉翻译时，原语中的某种词类往往不可能，也不一定需要译成汉语中的同一词类，而应根据译语的习惯和需要，转译成另一种词类。这就是词类的转译。

1. 英语名词的转译

1）转译成汉语的动词

A. 具有动作意义的名词的转译

英语中有一些具有动作意义的名词，这些名词汉译时往往要转译成汉语的动词，才会使译文连贯。如：.

supply chain

供应链

B. 某些动词加词根+ "er" 构成的名词，但它并不是真正表示某种职业，而是具有相应的动作意义时，可将其转译成汉语的动词。如：

He is the teacher of our logistics.

他教我们的物流课程。

C. 某些中心词是名词的习语化动词词组，其中心词可转译为动词。

D. 某些中心词是名词的介词短语，其中心词可转译为动词，如：at work

2) 转译成汉语的形容词

在英语中，有些表达事物性质的词语是由形容词派生而来的，其本身就具有形容词的性质。汉译这类词时，可转译为汉语的形容词。

2. 英语形容词的转译

1) 转译为汉语的动词

当形容词做表语、主语、补足语或其他成分时，可转译成汉语的动词。如：

If low-cost becomes available from supplier...

如果能从供应商获得低成本……

2) 转译为汉语的名词

当形容词做表语、主语、补足语或其他成分时，可转译成汉语的名词。

3. 英语动词的转译

无论是状态动词还是行为动词，在一定情况下均可转译为汉语的名词。

The instrument is characterized by its compactness and portability.

这种仪器的特点是结构紧凑和携带方便。

4. 英语中副词的转译

1) 转译成汉语的形容词

当英语中的动词转译成汉语的名词时，修饰动词的副词一般转译成形容词。

2) 由名词派生的副词

某些由名词派生的副词，可转译为汉语的名词。如：

Let us plot the electronic energy level vertically and assumed atomic spacing horizontally, as shown in Fig.4.

如图4所示，设垂直方向为电子能级，水平方向为假象的原子间距。

3) 转译为汉语的动词

英语中的某些副词，如 on, up, in, off, out, over, behind, away 等，在与系动词 be 构成合成谓语时，或作宾语补足语，或作状语，可转译为汉语的动词。如：

You may feel its electric charge flow through your body and away into the earth.

你会感觉到电流从你的身体中流过，然后离开而进入地下。

5. 英语中介词的转译

英语中不少介词，如 about, of, for, by, in, past, with, over, into, around, toward, through 等，都具有较强的动词意味，汉译时也可转译成汉语的动词。

机械制造类专业英语

Scene Two　Electrical Machining

Before you step into this scene, please learn by heart the following words and expressions with your classmates by looking up the dictionary or surfing the internet.

Look at the following words we may use in this scene and try to make out their Chinese meaning in a given time, whoever finishes off it first will be the winner.

due to	come about	space-age	electrochemical
unique	odd	dissimilar	time-consuming

_____

Now look at the following Chinese meaning with your colleagues carefully and try your best to make out their English meaning, finally dictate them for each other among your team workers.

电化学磨削　　　电化学加工　　　……所独有的　　　相比之下，相反，可是

_____

Most electrical machining processes have come about due to the development of apace-age metal alloys that are tough and hard to machine. Grinding cutting, and shaping with cutting tools are exacting, time-consuming, and expensive processes on such alloys. In some cases, machining with conventional equipment is almost impossible. A material like tungsten carbide, for example, is very difficult to machine even with diamond cutting tools.

Read the paragraph of the materials above and do your best to answer the following questions with the help of your classmates.

（1）Why do most electrical machining processes have come about?
_____

（2）Why some apace-age metal alloys can't be machined by grinding cutting, and shaping with cutting tools?
_____

（3）What kind of material can't be machined through conventional equipment even with diamond cutting tools?
_____

Read through the information above again with your workmates and try to fill in the blanks in the following short passage.

Most electrical machining processes have come about _____ the development of apace-age metal alloys that are _____ and _____ to machine. Grinding cutting, and shaping with cutting tools are exacting, _____, and expensive processes

Module Two Mechanical Machining

on such alloys. In some cases, machining with conventional equipment is almost impossible. A material like tungsten carbide, for example, is very difficult to machine _____ with diamond cutting tools.

 Read the materials above again with your workmates and then try to translate them into Chinese, and then discuss with your workmates again.

(1) Most electrical machining processes have come about due to the development of apace-age metal alloys.

_____

(2) Grinding cutting, and shaping with cutting tools are exacting, time-consuming, and expensive processes on such alloys.

_____

(3) A material like tungsten carbide is very difficult to machine even with diamond cutting tools.

_____

 On standard machine tools, electrical energy is converted by an electric motor into motion that is transmitted through a series of shafts, pulleys, belts, and gears to a cutting tool. In contrast, the electrical machining processes make direct use of electrical energy. This results in certain abilities that are unique to these processes, machining without tool rotation or without touching the work piece, machining odd shapes, and machining dissimilar metals without loading the tool.

Three major electrical machining processes are EDM (Electrical Discharge Machining), ECM (Electro Chemical Machining), and ECG (Electro Chemical Grinding).

 Read the above materials first with your colleagues and then try your best to give out the definition of electrical machining.

_____

 Work with your group members and read the information above again and try your best to translate the following terms into Chinese. Then dictate them among your group members.

work piece	electrical machining processes	gears
electrical energy	electric motor	shafts
cutting tool	belts	pulleys
odd shapes	loading the tool	

_____

 Survey your workplace and observe carefully, make out the process of electrical machining and ask the master for help if any problems, write down your experiences in your workplace.

_____

Work with your workmates reading the materials above carefully and then try to understand the kinds of processes of electrical machining. Please cooperate with your partners and write down your answers.

___

Read the paragraph of the materials above and do your best to answer the following questions with the help of your classmates.

(1) When electrical energy is converted by an electric motor into motion, what kinds of tools can be used to transmit to a cutting tool?

___

(2) Can you tell the differences between conventional machining and electrical machining?

___

Read the information above again with your roommates and also observe your workplace carefully and then write down the advantages of electrical machining?

___

 Practice spoken English.

Mr.Ma: How do you do? Welcome to visit our products. What kinds of equipment are you most interested in?

Mr. Li: I am interested in your NC machine tool. Would you give me a manual?

Mr. Ma: OK. May I know your name, please?

Mr. Li: Yes. Here is my calling card.

Mr. Ma: This is my card. Thank you for coming.

Mr. Li: Would you like to introduce this kind of NC machine tool?

Mr. Ma: No problem. Established in 1950, our company has a long history and abundant experience in manufacturing NC machine tools. We are highly praised by customers because of high quality of our products and our excellent after-sales services. Our products are widely sold in the world.

Mr. Li: Would you give me your quotation?

Mr. Ma: Let me think over, one hundred and twenty thousand RMB.

Mr. Li: Oh, your quotation is on the high side. I would like to visit other companies and then talk the price with you. See you later.

Mr. Ma: Hope to see you again.

New words and expressions we learn (let the students copy the new words and expressions he or she doesn't know in this scene and then dictate them among his/her group members).

___

Grammar 4　翻译的基本方式及其选择

翻译的基本方式有直译、意译、音译和阐译 4 种。

1．直译

所谓直译，就是既忠实于原文内容，又忠实于原文形式的翻译方式。

英汉两种语言句子主干成分的语序在基本句型中是一致的，这便为直译这种翻译方式奠定了基础。如英语中：

logistics deals with satisfying the customer.

译成汉语便是：物流的任务是为了满足客户的需求。

直译是能最全面、直接地完成翻译所承担的任务、达到公认的翻译标准的一种翻译方式。因此，直译是最为重要的翻译方式，也是所有翻译方式的基础。但应该指出的是，直译并不等于死译，也不能望文生义。否则就会出错、闹笑话。

2．意译

所谓意译，就是忠实于原文的内容，不拘泥于原文形式的一种翻译方式。

英汉两种语言毕竟分属两种迥然不同的语系，不可能在语序、句式和句子排列上完全契合。因此，在很多情况下，应该转换原文的表达方式，摆脱原文表层形式的束缚，按汉语的表达规则和习惯来遣词造句。也就是说，当译文与原文在神似与形似不能统一时，就只能重神似而不重形似了，这便是意译。

意译能在很大程度上，有效处理直译所不能解决的翻译问题，因此也是一种常用的翻译方式。采用意译方式时，应注意的是，一定要以神似为原则，不能意译过头。

3．音译

所谓音译，就是在译文中选择与原文发音相同的词语造句。如 laser 镭射，model 模特等，都是音译的例子。最早，中国没有"镭射"、"模特"这几类东西，当它们传入中国后，只能用音译的方式创造新词来表达。

4．阐译

所谓阐译，就是用解释性的语言来表述译文的一种翻译的方式。

阐译这种翻译方式，主要用于当两种语言之间没有完全对应的事物，但又不能简单用音译来翻译的情况。

Scene Three　Lathe

Before you step into this scene, please learn the following words and expressions by heart with your classmates by looking up the dictionary or surfing the internet.

机械制造类专业英语

Look at the following words we may use in this scene and match the English and Chinese meaning in a given time, whoever finishes off it first will be the winner.

drilling		n.	钻削
miling		n.	铣削
fasten		v.	固定，加固，连接
feed		n.	进给，送给
cutter		n.	刀具
clamp		n.	夹子
carriage		n.	溜板
bed		n.	床
headstock		n.	主轴箱
tailstock		n.	尾架
slide		v.	滑动
metalwork		n.	金属制品
spindle		n.	主轴
wobble		v.	摇晃
alignment		n.	队列
micrometer		n.	千分尺

The following expressions are useful. Before we finish off the task, please make out their English meaning with the help of your teammates or dictionary or the internet.

机床　　　　　　　铣床　　　　　　　刀座
主轴顶尖　　　　　尾架顶尖

___

A lathe is a machine tool for cutting metal from the surface of a round work fastened between the two lathe centers and turning around its axis. In turning the work a cutter moves in the direction parallel to the axis of rotation of the work or at an angle to this axis, cutting off the metal from the surface of the work. This movement of the cutter is called the feed. The cutter is clamped in the surface of the tool post which is mounted on the carriage. The carriage is the mechanism feeding the cutter in the needed direction. The lathe hand may feed the cutter by hand or may make it be fed automatically by means of special gears.

The largest part of the lathe is called the bed on which the headstock and the tailstock are fastened at opposite ends. On the upper part of the bed there are special ways upon which the carriage and the tailstock slide.

The two lathe centers are mounted in two spindles: one is held in the headstock spindle, while the other (the head center) is in the tailstock spindle.

Module Two Mechanical Machining

Read the paragraph of the materials above and do your best to answer the following questions with the help of your classmates.

(1) Where is the work piece fastened in a lathe?

_____

(2) What is feed according to the passage?

_____

(3) Where are the two lathe centers?

_____

Work with your group members and read the information above again and try your best to solve the following problems.

Please list the construction of a lathe with the help of your workmates.

_____

Read the materials above again with your workmates and then try to translate them into Chinese, and then discuss with your workmates.

(1) A lathe is a machine tool for cutting metal fastened between the two lathe centers and turning around its axis.

_____

(2) The cutter is clamped in the surface of the tool post which is mounted on the carriage.

_____

(3) The largest part of the lathe is called the bed on which the headstock and the tailstock are fastened at opposite ends.

_____

Read the materials above again with your workmates and then try to translate them into Chinese, and then discuss with your workmates.

lathe	round work	axis	turning the work
parallel to the axis	angle to this axis	cutter	feed
clamped	tool post	carriage	lathe hand

Survey your workplace and observe carefully, and then ask the master for help if there are any problems, write down your experiences in your workplace.

_____

The lathe chuck is used for chucking the work, which is used for clamping it so that it will rotate without wobbling while turning the chuck, usually, mounted on the headstock spindle may have different sizes and construction. If the work is perfectly round, it may be chucked in the

55

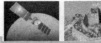

so-called three-jaw universal chuck, all the jaws of which are moved to the center by turning the screw. But if the work is not perfectly round, the four-jaw independent chuck should be used.

In turning different materials and works of different diameters, lathes must be run at different speeds. The gearbox contained in the headstock makes it possible to run the lathe at various speeds.

Before turning a work the lathe, the lathe center is to be alined: that means that the axes of both centers must be on one line.

The alignment of the lathe centers may be tested by taking a cut and then measuring both ends of the cut with a micrometer.

Not all works should be fastened between the two centers of the lathe. A short work may be turned without using the dead center, by simply chucking it properly at the spindle of the headstock.

Discuss with your colleagues and try to share the your ideas and then decide which of the following expression is right or wrong. T for Right and F for wrong.

Tick the correct response in the relevant box.

（1）The lathe chuck is used for chucking the work for rotation without wobbling while turning the chuck.　　　　　　　　　　　　　　　　　　　　　　　　　　　　　　　　　（　　）

（2）All the jaws of three-jaw universal chuck are moved to the centre by turning the screw.
　　　　　　　　　　　　　　　　　　　　　　　　　　　　　　　　　　　　　（　　）

（3）The jaws of the four-jaw chuck are moved to the center by turning the screw independently.
　　　　　　　　　　　　　　　　　　　　　　　　　　　　　　　　　　　　　（　　）

Read through the information above again with your workmates and try to fill in the blanks in the following short passage.

In turning _____ materials and works of _____ diameters, lathes must be run at _____ speeds. The gearbox contained in the headstock makes it possible to run the lathe _____ various speeds. Before turning a work, the lathe center is to be _____: that means that the axes of both centers must be on _____ line. The alignment of the lathe centers may be tested by taking a cut and then measuring both ends of the cut with a _____. Not _____ works should be fastened between the two centers of the lathe. A short work may be turned _____ using the dead center, by simply chucking it properly at the spindle of the headstock.

Read the materials above again with your workmates and then try to translate them into Chinese, and then discuss with your workmates.

lathe chuck	clamping	wobbling	turning the chuck
headstock spindle	construction	gearbox	short work
three-jaw universal chuck	four-jaw chuck	lathe center	dead center

Module Two Mechanical Machining

Read the materials above again and then do your best to make out the Chinese meaning of the following sentences with the help of your teammates.

(1) The lathe chuck is used for chucking the work, which is for clamping it so that it will rotate without wobbling while turning the chuck.

_____

(2) In turning different materials and works of different diameters, lathes must be run at different speeds.

_____

(3) Not all works should be fastened between the two centers of the lathe.

_____

Survey your workplace and observe carefully, make out the importance of lathe in your workplace and then ask the master for help if there are any problems. Compare the differences of the chucks.

	three-jaw universal chuck	four-jaw chuck
definition		
applied parts		
advantages		
disadvantages		

Survey your workplace again with your classmates and observe carefully, then write down the meaning of the following figure and translate relative Chinese words into English.

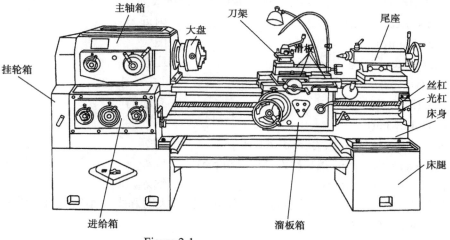

Figure 2-1 _____

Practice spoken English.

Mr. Cook: Excuse me, may I ask you a question, Jay?

Jay: Of course.

Mr. Cook: Do you know about the structure of the lathe?

Jay: Yes, I know some of them.

Mr. Cook: What's that?

Jay: It's a headstock. Sometimes people call it the heart of a lathe.

Mr. Cook: What's this then?

Jay: It's a chuck and a work-piece can be mounted on it.

Mr. Cook: What's that?

Jay: It's tool post and carriage. The cutting tools are put here.

Mr. Cook: What's that?

Jay: It's center and tailstock.

Mr. Cook: What's that?

Jay: It is a bed. These are all the main structure of the lathe. Are those clear to you?

Mr. Cook: Yes, you are very kind.

Jay: Delighted to have been of assistance.

New words and expressions we learn（let the students copy the new words and expressions he or she doesn't know in this scene and then dictate them among his/her group members）.

Grammar 5　缩略词的解释

缩略词在这里主要是指一个单词（也有两个或两个以上的）缩略而成的词，大多均以一个句号结束（节略词无句号），英语中缩写和节略词数量很大，有的甚至长达10个字母，如：

  Appx. (Appendix)　　　　　　　　　　　附近
  Fig. (Figure)　　　　　　　　　　　　　图
  Cf. (confer [拉]意为 compare)　　　　　比较
  Vs. (versus)　　　　　　　　　　　　　与……比较
  Ltd. (limited)　　　　　　　　　　　　　有限的
  ff. (following)　　　　　　　　　　　　　下列
  etc. (et cetera [拉] 意为 and so forth)　　等等
  e.g. (exempli gratia [拉] 意为 for example)　例如
  I. O. (Input Output)　　　　　　　　　　输入，输出

1. 缩写和节略较为灵活，以小写字母为主，也有大写和大、小写混合的缩写和节略。具体有如下几种缩略方式。

（1）取首字母

  Lab. (laboratory)   实验室

  Freq. (frequency)   频率

  Ent. (entry)   入口

  Dyn. (dynamometer)   功率计

  Sig. (signal)   信号

或取前两个或几个音节

  Navig. (navigation)   导航

  Telecom. (teleconference)   远程电视电话会议

（2）取主要字母（以辅音字母为主）

  Txt. (text)   电文

  Tty. (teletype)   电传打字

  St. (subsequent)   随后的

  ss. (supersensitive)   超灵敏度

  gr. (ground)   接地

  chg. (charge)   电荷

过去分词常连带取词尾

  DISCONTD (discontinued)   终止

  DLVD （delivered）   已投送的

（3）取读音

  Fax (facsimile)   传真

  Ur (your)   你的，你们的

  BOZ (both)   双方

  WIZ (with)   和……一起

  Ky (key)   开关

（4）取主要字母和音节（尾字母居多）

  Apx (appendix)   附录

  Assy (assembly)   装备

  Desr (designer)   设计员

  Mbar (millibar)   毫巴

  Msec (millisecond)   毫秒

2．缩略词语汉译的一般顺序和技巧要点

（1）根据原文所给全文（大都附在该缩略词语后的括弧内）解释汉译；

（2）根据缩略词语构成规律写出全文，然后汉译；

（3）部分单词缩略可先尝试读出再汉译，如：

  FXD——fixed

  RAX——remarks

  WUD——would

（4）查阅相关工具书。

3．在上述诸法都无效的情况下，可尝试若干模糊技巧

（1）确定该缩略词语所在文章所属的专业范围，范围越小越好。在许多缩略词语工具书上，同一缩略语往往有几个解释，缩小专业范围有助于正确选择，如：

```
PCM    plug-compatible mainframe        插接兼容主机
       Process control monitor           过程控制监视器
       Pulse code modulation             脉冲编码调制
       Punched card machine              穿孔卡片机
```

究竟选择哪一个，要根据专业内容决定。

（2）利用缩略词语出现的上下文猜破。从上下文有时可以找到缩略词语的全文，通过文章内容的提示也可对缩略词语做出较准确的推测。

（3）分解拼凑。有些缩略词语可能由两个或多个缩略词语组成。先对其施加分割，得到意译后再合在一起，如

DVTLC=DIG+LC（direct-coupling transistor+logic circuit）
直接耦合晶体管逻辑电路
DIG-MOD=DIG+MOD（digital+modulation）
数字调制器
FACTS=FAC+TS（facsimile+transmission system）
传真传输系统
FAXAMMODE=FAX+AM+MO+DE（facsimile+amplitude+modulation+mode）
传真调幅方式
TELECOMM=TELE+COMM（telephone+communication）

Scene Four   Milling

Before you step into this scene, please learn the following words and expressions by heart with your classmates by looking up the dictionary or surfing the internet.

Look at the following words we may use in this scene and match the English and Chinese meaning in a given time, whoever finishes it off first will be the winner.

remove		v. 旋转
mount		v. 固定
arbor		n. 心轴
spacer		n. 垫片
bushing		n. 衬套
yoke		n. 刀杆支架
slit		v. 切开

Module Two　Mechanical Machining

clockwise	*a.* 顺时针
regular	*a.* 规则的
particular	*a.* 特别的
horizontal	*a.* 水平的
various	*a.* 不同点
variation	*n.* 变化
knee	*n.* 升降台
universal	*a.* 万能的
elevate	*v.* 升高
index	*n.* 分度头
stock	*n.* 台

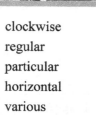 The following expressions are useful. Before we finish off the task, please make out their English meaning with the help of your teammates or dictionary or the internet.

装有　　　　　　　分类为　　　　　　　分成
根据　　　　　　　从……出去　　　　　虎钳
装置　　　　　　　横向的

The milling machine is a machine that removes metal from the work with a revolving milling cutter as the work is fed against it, the milling cutter is mounted on an arbor where it is held in place by spacers or bushings, the arbor is fixed in the spindle with one end, while the other end of the arbor rotates in the bearing mounted on the arbor yoke.

The milling cutters are generally made from high speed steel and available in different sizes and shapes. there are such kinds of milling cutters as cylindrical cutters, end milling cutters(for face milling), form milling cutters, angular cutters, side and face cutters, skate ting saw, etc. these cutters may differ in the direction of their operation i.e. they may cut revolving either clockwise or counter-clockwise.

Regular or irregular shaped work may be produced on a milling machine. Designs varying according to the particular class of work wanted. According to the position of the spindle, the milling machines may be divided into two groups of vertical spindle milling machines and horizontal spindle milling machines. Milling machines may be grouped into various classes.

According to variation in general design as the "column and knee type", the manufacturing types, and the planer type of milling machine. According to the table design, the milling machines may be classified as universal and plain milling machines.

Read the paragraph of the materials above and do your best to answer the following questions with the help of your classmates.

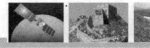

（1）How many kinds of milling cutters?
_____
（2）What is the milling machine?
_____
（4）How many kinds of milling machines according to general design?
_____
（5）How many kinds of milling machines according to the position of the spindle?
_____

Work with your group members and read the information above again and try your best to solve the following problems.

Write down the types of milling machines according to different criterion.

general design	
position of the spindle	
table design	

Read the materials above again with your workmates and then discuss with your colleagues and fill in the following table about the types of milling cutters. Tick it out if it is.

cylindrical cutters （ ） end milling cutters （ ） side and face cutters （ ）
form milling cutters （ ） angular cutters （ ） slitting saw （ ）

Survey your workplace again with your classmates and observe carefully then write down the meaning of the following figure. Dictate the new terminologies for each other among your group members.

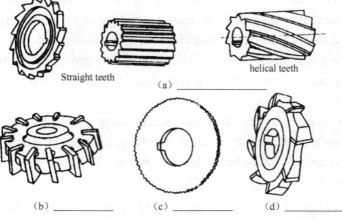

Straight teeth   helical teeth
(a) _____

(b) _____ (c) _____ (d) _____

Figure 2-2 _____

Module Two Mechanical Machining

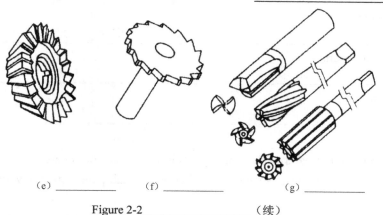

(e) _____   (f) _____   (g) _____

Figure 2-2 _____ （续）

Read the materials above again with your workmates and then try to translate them into Chinese, and finally discuss with your workmates.

(1) The milling cutter is mounted on an arbor where it is held in place by spacers or bushings.

_____

(2) The milling cutters are generally made from high speed steel and available in different sizes and shapes.

_____

(3) Designs varying according to the particular class of work wanted.

_____

Read through the information above again with your workmates and try to fill in the blanks in the following short passage.

According to the _____ of the spindle, the milling machines may be divided into _____ groups of vertical spindle milling machines and _____ spindle milling machines. Milling machines may be grouped into various _____.

According to variation in general _____ as the "column and knee type", the _____ types, and the _____ type of milling machine. According to the _____ design, the milling machines may be classified as universal and _____ milling machines.

Read through the information above again with your workmates and try to put them into the relative place about the types of milling machines, put them into different classification.

① vertical spindle milling machines　　② universal milling machines
③ plain milling machines　　④ the planer type of milling machine
⑤ horizontal spindle milling machines　　⑥ the manufacturing types
⑦ column and knee type

General design
_____

Position of the spindle
_____

Table design
_____

☞ The most impartment parts of the milling machine are starting levers, spindle, column, knee, elevating screw, table, index head, speed levers, feed levers, table movement levers, foot stock, and arbor yoke.

The spindle of the milling machine is driven by an electric motor through a train of gears mounted in the column. The table of the plain milling machine may travel only at right angles to the spindle, while the universal milling machine is provided with a table that may be swiveled on the transverse slide for milling gear teeth, threads etc.

Various attachments are used for increasing the range of work that can be performed on a milling machine.

The dividing head (also called an index head) is a device used to divide the periphery of a piece work into any number of equal parts, and to hold the work in the required position while the cuts are being made.

Various kinds of vises may be used for holding the work in a milling machine, the most common being the plain vise, and the swivel vise.

☞ Read the materials above again with your workmates and then try to translate the following phrases into Chinese, and then discuss with your workmates.

starting levers	spindle	column	knee
elevating screw	table	index head	speed levers
feed levers	table movement levers	foot stock	arbor yoke

☞ Read the materials above again and then do your best to make out the Chinese meaning of the following sentences with the help of your teammates.

(1) The spindle of the milling machine is driven by an electric motor through a train of gears mounted in the column.

(2) Various attachments are used for increasing the range of work.

(3) The dividing head is a device used to divide the periphery of a piece work into any number of equal parts.

Module Two Mechanical Machining

⚠ Discuss with your colleagues and try to share your ideas and then decide which of the following expression is right or wrong. T for Right and F for wrong.

Tick the correct response in the relevant box.

(1) The table of the plain milling machine may travel only at right angles to the spindle.
( )
(2) The table of universal milling machine may be swiveled on the transverse skatede. ( )
(3) The plain vise and the swivel vise are common vise in milling machine. ( )

Survey your workplace and observe carefully, make out the importance of milling machines in your workplace and then ask the master for help if there are any problems, write down your experiences in your workplace.

Read through the information above again with your workmates and then tick out the following items which belong to the parts of a milling machine.

starting levers	( )	column	( )
elevating screw	( )	index head	( )
feed levers	( )	foot stock	( )
spindle	( )	table	( )
table movement levers	( )	knee	( )
speed levers	( )	arbor yoke	( )

Read through the above information again with your workmates and try to fill in the blanks in the following short passage.

Various attachments are used for increasing the _____ of work that can be performed on a _____ machine. The dividing head (also called an _____) is a device used to divide the periphery of a piece work into any number of _____ parts, and to hold the work in the _____ position while the cuts are being made. Various kinds of _____ may be used for holding the work in a milling machine, the most common being the _____ vise, and the _____ vise.

Survey your workplace again with your classmates and observe carefully, then write down the meaning of the following figure.

65

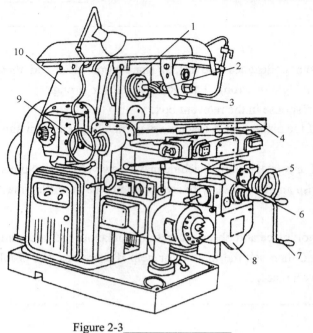

Figure 2-3 _____

 Survey your workplace again with your classmates and observe carefully, then write down the meaning of the following figures. Dictate the new terminologies for each other among your group members.

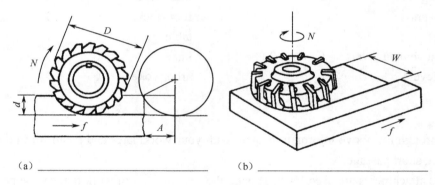

(a) _____　　　　　　(b) _____

Figure 2-4 _____

Practice spoken English.

    Jay: How do you do! My name is Jay.
    Miss Kate: How do you do!
    Jay: Are there any vacancies in your factory?
    Miss Kate: Yes. We need some new workers. What is your major?
    Jay: I major in mechanical manufacturing, and I think I am good at operating milling machine.
    Miss Kate: What sort of job do you want to do?

Module Two Mechanical Machining

Jay: I would like to be a milling worker. I think I can do my best to it.
Miss Kate: Great! What time can you start to work?
Jay: Anytime.
Miss Kate: Very nice. How much do you expect in salary?
Jay: 700 yuan.
Miss Kate: OK! Please fill in this form. If anything turns, I'll contact with you.
Jay: Thank you!
Miss Kate: You're welcome.

New words and expressions we learn（let the students copy the new words and expressions he or she doesn't know in this scene and then dictate them among his/her group members）.

Grammar 6　短语的翻译

1．分词短语的译法

由于状语性分词短语一般并不表示强调，且与主干之间一般并无表示逻辑关系的连接词，因此，翻译含有此类短语的句子，其关键是根据上下文的含义，弄清枝干之间的逻辑关系。

1）现在分词短语的译法

A．译成动词并列式分句或后续分句

所谓译成动词并列式分句或后续分句，就是在一个主语下使用两个并列动词，其中一个是原文中的主句的谓语动词，另一个是分词短语中的分词原形。两并列式分句之间用逗号分开。

动词并列式分句最基本的特性是表示主要意义的分句与表示次要意义的分句共同用一个主语，形成连动关系。

Being important and clear, the environment are useful for human being.

B．译成复合式句式

所谓译成复合式分句，就是将现在分词短语译成独立的句子，并且使它与原主句的译句一同构成有逻辑关系的并列或从属句式。如：

The man is called national hero, with a strong and brave heart.

他拥有一颗坚强而勇敢的心，所以称其为民族英雄。

C．译成包孕式

此处所说包孕，是指将状语性分词短语译成包孕于句中的适当成分，在译文中形成句中句的形式。如：

The inventory management is very useful for logistic, including that of transpotation.

存货管理在物流中非常重要，包括运输。

D．译成外位语句

如果句子太长，或成分与层次较多，有时需将其切分。这时，常常用一个指代成分来代替先行部分。指代词前的先行部分就被称为外位语，指代词本身是本位语。本位语承接外位语，形成明晰的层次，与汉语的行文习惯一致。

E．译成前置无主句

所谓前置无主句，实际上就是连动式分句。其前句与后句仍共用一个主语，只是前句中的主语不出现，主语仅出现在后句中。

2）过去分词短语的译法

A．译成定语修饰词

有些表时间、地点或方式的状语性过去分词短语，有时可译为定语修饰词。如：

Bred in the areas around the equator, this protozoan proves very inactive out of the tropical zone.

这种在赤道周围繁殖的原生物，在热带地区以外的地方是极不活跃的。

B．译成动词并列式分句或后续分句

The road of the street is in stones, called stone street.

这条道路是用石头砌成的，因此叫石头街。

C．译成独立句

He was Michael Faraday, born in Yorkshire, 1791.

这人便是法拉第，他于1791年出生于约克郡。

2．介词短语的译法

1）做定语的介词短语译成前置附加语

做定语的介词短语一般后置，而汉语的定语均为前置，因此应将其译成前置附加语。

2）做状语的介词短语的译法

A．译成单词型状语

B．译成短语型状语

C．译成并列句

D．译成动词并列式分词

E．译成复句

Scene Five　Radial Drilling Machines

Before you step into this scene, please learn the following words and expressions by heart with your classmates by looking up the dictionary or surfing the internet.

Look at the following words we may use in this scene and match the English and Chinese meaning in a given time, whoever finishes it off first will be the winner.

radial	*adj.* 半径的
radially	*adv.* 沿径向
handle	*n.* 把手

Module Two Mechanical Machining

arm	n. 臂
circle	n. 圆
control	n. 控制
select	vt. 选择
selector	n. 选择器
lever	n. 杠杆
engage	n. 接合
deal	vt. 量

The following expressions may be useful. Before we finish off the task, please make out their English meaning with the help of your teammates or dictionary or surfing the internet.

摇臂钻床　　　　　　摇臂　　　　　　　选速手柄
大量的

The radial drilling machine is designed for handling large work-pieces that cannot be easily moved. The drilling machine head is mounted on a heavy radial arm which may be from three to twelve or more feet long. This arm can be raised or lowered with power and can be turned in a complete circle around the column. The drilling head moves back and forth along this arm. On most radial drilling machines, movement of the arm, drill head, and spindle is controlled by power.

Read the paragraph of the materials above and do your best to answer the following questions with the help of your classmates.

（1）What is the radial drilling machine designed for?
_____

（2）Where is the drilling machine head mounted?
_____

（3）What controlled the movement of the arm, drill head, and spindle?
_____

Work with your group members and read the information above again and try your best to solve the following problems.

Please write down the function of the radial drilling machine.
_____

Read the materials above again with your workmates and then try to translate the following sentences about the function of radial drilling machine into Chinese, and finally discuss with your workmates.

(1) The radial drilling machine is designed for handling large work-pieces that cannot be easily moved.

_____

(2) The drilling head moves back and forth along this arm.

_____

(3) This arm can be raised or lowered with power and can be turned in a complete circle around the column.

_____

 Work with your classmates trying to translate the following phrases into Chinese.

the radial drilling machine   the drilling machine head
a heavy radial arm   raised or lowered with power
moves back and forth   movement of the arm, drill head, and spindle

_____

Survey your workplace and observe carefully, and then ask the master for help if there are any problems, write down your experiences in your workplace.

_____

Spindle feeds and speeds are controlled by selector levers which engage the proper gears in the drill head. Depth of feed is also controlled directly in the drill head by a suitable mechanism. In addition to drills, other tools such as reamers and boring heads can also be used, these tools add a great deal of versatility to this machine.

 Work with your classmates trying to translate the following phrases into English.

进给量和转速   选速手柄
主轴箱   齿轮
进给深度   适当的机构
钻头   刀具
铰刀   镗头

_____

Discuss with your colleagues and try to share your ideas and then make sure which of the following expression is right or wrong. T for Right and F for wrong.

Tick the correct response in the relevant box.

(1) Spindle feeds and speeds are controlled by selector levers which engage the proper gears in the drill head.   (　　)

(2) Depth of feed is also controlled directly in the drill head by a suitable mechanism. (   )

(3) the jaws of the four-jaw chuck are moved to the center by turning the screw independently.

(   )

Read through the information above again with your workmates and try to fill in the blanks in the following short passage.

Spindle feeds and speeds are controlled by selector _____ which engage the proper gears in the drill head. Depth of _____ is also controlled directly in the drill head by a _____ mechanism. In addition to drills, other tools such as _____ and _____ can also be used; these tools add a great deal of _____ to this machine.

Survey your workplace and observe carefully, make out the importance of radial drilling machines in your workplace and then ask the master for help if any problems.

Practice spoken English.

Mr. Sun: How do you do? Welcome to visit our products. What kinds of equipment are you most interested in?

Mr. Li: I am interested in your radial drilling machines. Would you give me a manual of Model XK 5025?

Mr. Sun: OK, may I know your name, please?

Mr. Li: Yes, here is my card.

Mr. Sun: This is my card. Thank you for coming.

Mr. Li: Would you like to introduce this kind of radial drilling machine?

Mr. Sun: No problem. Established in 1960, our company has a long history and abundant experience in radial drilling machine. We are highly praised by customers because of high quality of our products and our excellent after-sales services. Our products are widely sold in the world.

Mr. Sun: Would you give me your quotation?

Mr. Sun: Let me think over, 120 000RMB.

Mr. Li: Oh, your quotation is on the high side. I would like to visit other company and then talk the price with you. See you later.

Mr. Sun: Hope to see you again.

New words and expressions we learn (let the students copy the new words and expressions he or she doesn't know in this scene and then dictate them among his/her group members).

Grammar 7　分词

分词是非限定动词的一中,分词有现在分词和过去分词两类。

1. 形式及特征

现在分词由动词原型+"ing"构成。规则动词的过去分词由动词原形+"ed"构成。不规则动词的过去分词没有规则。

分词有形容词和动词特征,可以有宾语或状语,现在分词有一般式和完成式,有主动语态和被动语态,而过去分词只有一般式,没有完成式,也没有主动语态。

2. 用法

1)定语

作定语用的分词如果是单词,则放在它所修饰的名词之前,如:

China is a developing country.

中国是一个发展中国家。

It is a well-written article.

这是一篇好文章。

作定语用的分词如果是分词短语,则放在它所修饰的名词之后,相当于定语从句。

Debts owed to creditors are known as liabilities.

欠债权人的债叫做负债

The man talking with the customers is our accountant.

与顾客们谈话的人是我们的会计。

2)作表语

This novel is very moving.

这本小说很动人。

The theory sounds quite convincing.

这种说法听起来很有说服力。

分词作表语时,要与现在进行时与被动语态区分开,如:

He is reading the novel.(现在进行时)

他正在读小说

His interest is reading.(现在分词作表语)

他的兴趣是阅读

The report is well written.(过去分词作表语)

报告写得很好

The report was written by an officer.(被动语态)

报告是一个官员写的。

3)作状语

(1)表示时间

Having worked for three hours, she took a rest.

Module Two Mechanical Machining

工作了三个小时之后,她休息了一会儿。
Opening the drawer he took out a book.
打开抽屉,他拿出一本书。
(2) 表示原因
Being debits always recorded in amounts equal to credits, the debits and credits should always equal each other.
由于借贷所记的金额是相等的,所以借与贷必须相等。
Inspired by the teacher, the students studied even harder.
在老师的鼓励下,学生们更加努力地学习。
(3) 表示伴随或方式
Laughing and talking, the children went out into the garden.
孩子们又说又笑地走进花园。
4) 作宾语补足语
We must get everything straightened out.
我们必须把一切弄清楚。
I found that store almost completely rebuilt.
我发现那个商店几乎全部改建过了。

Scene Six Grinding Machines

Before you step into this scene, please learn the following words and expressions by heart with your classmates by looking up the dictionary or surfing the internet.

Look at the following words we may use in this scene and match the English and Chinese meaning.

conical	*adj.* 圆锥的
surplus	*n.* 过剩
exterior	*adj.* 外面的
taper	*n.* 锥形
sharpen	*vt.* 使锐利
substitute	*n.* 代替者
base	*n.* 床身
concentricity	*n.* 同轴度
periphery	*n.* 圆柱表面
infeed	*n.* 横向进给
interior	*adj.* 内部的

The following expressions are useful. Before we finish off the task, please make out their English meaning with the help of your teammates or dictionary or the internet.

单人车刀	成形磨削
横向磨削	内圆磨头
圆柱孔	万能磨床
圆柱体外表面	横向全面进给磨削
内圆磨床	平面磨床
无心磨床	外圆磨床
工具磨床	磨床
余量	

A grinding machine is a machine which employs a grinding wheel for producing cylindrical, conical or plane surfaces accurately and economically and to the proper shape, size, and finish. The surplus stock is removed by feeding the work against the revolving wheel or by forcing the revolving wheel against the work.

There is a great variety of grinding machine. The machines that are generally used are cutter grinder, surface grinder, center less grinder, external grinder, internal grinder, and tool grinder. These machines are usually classified as to size by the largest piece of work which they can be completely machined (as 6×18 inch external grinder).

Five types of grinding operations are performed on grinding machines.

(1) Cylindrical grinding. Exterior cylinders and tapers are ground on plain grinding machines and on center less grinders.

The construction of universal grinding machines is similar to that of plain grinders. They are equipped with an internal grinding attachment, however, no that they can do both cylindrical and internal grinding.

(2) Internal grinding. Interior cylinders and tapers are ground on internal grinding machines and on center less grinders.

(3) Surface grinding. Plane surface are ground on several types of surface-grinding machines.

(4) Tool and cutter grinding. Reamers, milling cutters and other tools are sharpened on tool and cutter grinding machines.

(5) Form grinding. Some machine tools such as thread, and gear grinders are designed especially for form grinding.

Survey your workplace and observe carefully, and then ask the master for help if there are any problems, write down your experiences in your workplace. Then write down the meaning of the following figures. Give each of the following figures a topic. Dictate the new words for each other among your group members.

Module Two Mechanical Machining

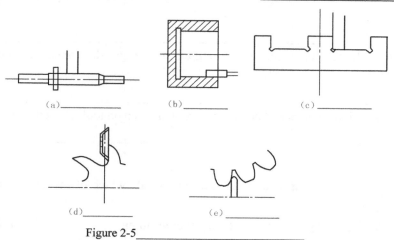

(a)_____ (b)_____ (c)_____

(d)_____ (e)_____

Figure 2-5_____

Read the paragraph of the materials above and do your best to answer the following questions with the help of your classmates.

(1) What is grinding machine?
_____

(2) How many types of grinding machine?
_____

(3) How many grinding operations are performed on grinding machines?
_____

Read through the above information again with your workmates and try to fill in the blanks in the following short passage. Dictate the words and expressions for each other among your roommates.

A grinding _____ is a machine which employs a grinding _____ for producing cylindrical, conical or plane _____ accurately and economically and to the _____ shape, size, and finish. The surplus stock is removed by feeding the work against the revolving wheel or by forcing the revolving wheel _____ the work.

There is a great variety _____ grinding machine. The machines that are generally used are _____ grinder, _____ grinder, _____ grinder, _____ grinder, _____ grinder, and _____ grinder. These machines are usually classified as to _____ by the largest piece of work which they can be completely machined.

Read the materials above again with your workmates and then try to translate the following sentences about grinding machine into Chinese, and finally discuss with your workmates.

(1) Exterior cylinders and tapers are ground on plain grinding machines and on center less

grinders.

(2) Interior cylinders and tapers are ground on internal grinding machines and on center less grinders.

(3) Reamers, milling cutters and other tools are sharpened on tool and cutter grinding machines.

Read the following materials carefully and then decide which of the definition belong to the corresponding place.

(1) cylindrical grinding　　　(2) internal grinding　　　(3) surface grinding
(4) tool and cutter grinding　　(5) form grinding

(　　) Reamers, milling cutters and other tools are sharpened on tool and cutter grinding machines.

(　　) Exterior cylinders and tapers are ground on plain grinding machines and on center less grinders.

(　　) Some machine tools such as thread, and gear grinders are designed especially for form grinding.

(　　) Interior cylinders and tapers are ground on internal grinding machines and on center less grinders.

(　　) Plane surface are ground on several types of surface-grinding machines.

Principal parts of a plain grinding machine are as follows:

(1) Base. The main casting of a plain grinding machine is a base that rests on the floor.

(2) Tables. A sliding table, which is mounted on ways at the front and top of the base, may be moved longitudinally by hand or power to feed work-pieces past the face of the grinding wheel. If the centerline of the swivel table is set at an angle, a taper will be ground.

(3) Headstock and tailstock. A motor-driver headstock and a tailstock are mounted on the left and right ends, respectively, of the swivel table for holding work-pieces on centers. The headstock center on a grinder is a dead center that is both centers are "dead" to insure concentricity of the periphery of groundwork with its axis.

(4) Wheel head. A wheel head carries a grinding wheel and its driving motor is mounted on a slider at the top and rear of the base. The wheel head may be moved perpendicularly to the table ways by hand or power, to feed the wheel to the work.

Either of two methods may be used to grind cylinders and tapers on plain grinding machines. In traverse-cut grinding, the work-piece is fed longitudinally past the face of the grinding wheel, and the wheel is fed toward the work after each reversal of the table until the piece is finished to

Module Two Mechanical Machining

size. The in feed at each reversal should be from 0.001 to 0.005 inch for roughing, and from 0.0005 to 0.0025 inch for finishing. in plunge-cut grinding, a grinding wheel that has a face width equal to or slightly longer than the surfaces being ground is employed and is fed transversely until the work-piece has the required diameter. No longitudinal feeding motion is used.

Survey your workplace and observe carefully, and then ask the master for help if there are any problems, write down your experiences in your workplace. Then write down the meaning of the following figure. Give a topic to the following figure. Dictate the new words for each other among your group members.

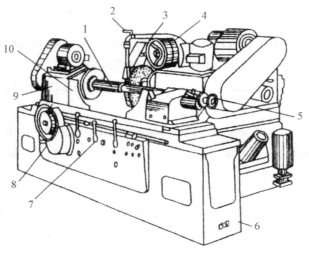

Figure 2-6 _____

The following expressions are useful. Before we finish off the task, please make out their Chinese meaning with the help of your teammates or dictionary or the internet.

plain grinding machine	base
tables	headstock and tailstock
wheel head	sliding table
grinding wheel	swivel table
centers	traverse-cut grinding
roughing	

Survey your workplace and observe carefully, and then ask the master for help if there are any problems, write down your experiences in your workplace. Please talk about the principal parts of a plain grinding machine and then filling in the following blanks.

parts of a plain grinding machine	base	
	tables	
	headstock and tailstock	
	wheel head	

Survey your workplace and observe carefully, and then ask the master for help if there are any problems, write down your experiences in your workplace.

_____

Translate the following sentences about a plain grinding machine with your cooperators.

(1) A plain grinding machine is essentially a lathe on which a grinding wheel has been substituted for the single-point tool.

_____

(2) The main casting of a plain grinding machine is a base that rests on the floor.

_____

(3) No longitudinal feeding motion is used.

_____

(4) Either of two methods may be used to grind cylinders and tapers on plain grinding machines.

_____

Discuss with your colleagues and try to share your ideas and then make sure which of the following expression is right or wrong. T for Right and F for wrong.

Tick the advantages of hydraulic machine tool drive.
(1) There are roughly four parts of a plain grinding machine.   (   )
(2) A lathe is somewhat like a plain grinding machine.   (   )
(3) The headstock centres are both "dead" centres.   (   )
(4) If the center line of the swivel table is set at an angle, a taper will be ground. (   )

Survey your workplace and observe carefully, make out the importance of grinding machines in your workplace and then ask the master for help if any problems, write down your experiences in your workplace.

Module Two Mechanical Machining

Look at the following words we may use in this scene and match the English and Chinese meaning.

grinder	n. 磨床
capable	adj. 有能力的
swivel	v. 旋转
degree	adv. 程度
line	n. 线
footstock	n. 尾架
dog	n. 挡块
rapid	adj. 迅速的，快的
rapidly	adv. 快地，迅速地
rate	n. 率，比率，速度
oppositely	adv. 在对面，相反地
direction	n. 方向
contact	v. 接触，触点
capable of	能……的，可以……的
center line	中心线
at the rate of	按……速率（比率）
in opposite direction	朝相反方向
in the direction of…	朝……方向
rpm=revolution per minute	转/分

Plain center-type cylindrical grinder is used to grind the outside of cylindrical parts. The table that is fastened to the bed of this machine is capable of back-and-forth (reciprocating) movement and can also be swiveled about 10 degrees on either side of the center line. Both the headstock and the footstock are mounted on this table. The grinding wheel is mounted on a spindle. It can be moved to and from the work-piece.

Both the grinding wheel and the work-piece rotate during the machining operations. The grinding wheel turns rapidly-825 to 1250 rpm. The work piece turns a much slower rate-100 to 400 rpm. The surfaces of the grinding wheel and the work piece move in opposite directions at their line of contact. The table is used to move the work-piece across the front of the wheel.

 Give a topic to the materials above.

机械制造类专业英语

 Practice spoken English.

Miss Jone: Hello, Mr.Christer, nice to meet you!

Mr. Christer: Hi Miss jone, nice to meet you, too.

Miss Jone: What are you doing?

Mr. Christer: I am checking the grinding machines.

Miss Jone: I know your major is engineering. Could you tell me how many kinds of grinding machines there are in common use?

Mr. Christer: There are …

Miss Jone: What are they?

Mr. Christer: They are …

Miss Jone: How do the grinding machines work?

Mr. Christer: Well …

Miss Jone: I'm really very grateful to you.

Mr. Christer: Not at all.

New words and expressions we learn（let the students copy the new words and expressions he or she doesn't know in this scene and then dictate them among his/her group members）.

Grammar 8 动词不定式

1. 概念

不定式是动词的一种非限定形式，是由不定式记号 to 加动词原型构成的。不定式具有动词的特征，即及物动词可带宾语，可被状语修饰，也可有语态的变化的。

当一个动词不定式与另一个连用时，不定式前面一般都有 to，如：

I want to go to the station.

我想去车站。

而在某些动词后面（如 can，shall，do，will，may，need），to 要省略，如：

You may do the work.

你可以做这项工作。

2. 用法

动词不定式在句子中可以作主语、宾语、标语、定语、状语，另外还可和名词一起构成复合宾语。

1）不定式作主语

To debit an account means to enter an amount on the left or debit side of the account.

借记一个账户，是指把金额记在账户的左方或借方。

有的情况下，可用 it 作形式主语来代替不定式做主语，如：

It is important to talk with her.
与她谈话是重要的。

It is common to use tabular arrangements rather than narrative-type reports.
通常使用表格式的排列 而不是用叙事型的报告。

2）不定式作宾语

The owner wanted to make a large purchase of gift before Christmas.
老板想在圣诞节前做一大笔礼品生意。

He decided to write a report at the end of the month.
他决定在月底写一个报告。

3）不定式作表语

The purpose of business accounting is to provide information about the current financial operations and condition of an enterprise to individuals agencies and organizations.
企业会计的目的是向个人、机构、组织提供企业当期财务经营状况的信息。

Her wish is to become an accountant.
她的愿望是成为一名会计师。

4）不定式作定语

Today, I have a lot of work to do.
今天我有许多事要做。

He has a meeting to attend this evening.
今晚他要去开个会。

It is time to have breakfast.
是吃早饭的时候了。

5）不定式作状语

（1）表示目的

He sat down to read newspapers.
他坐下来读报。

（2）表示原因

I'm very glad to hear the news.
我很高兴听到这个消息。

（3）This perfume is too expensive to purchase.
这瓶香水太贵，买不起。

6）复合宾语中的不定式

不定式可与名词或代词组成复合结构，作动词的宾语，即复合宾语。

可以有这样宾语的动词很多，如 ask, tell, compel, get, allow, wish, want, like, hate, intend, advise, order 等。

He wanted me to come back soon.
他希望我马上回来。

He asked the teacher to explain the sentences.
他要求老师解释这个句子。

注意：在 make，let，have，see，hear，watch，notice，feel 等词的复合宾语中，不定式须省略掉 to，而 help 后不定式的 to 可有可无，如：

I hear children sing this song.
我听到孩子们唱这首歌。

Scene Seven　　Power in the Workshop

Before you step into this scene, please learn the following words and expressions by heart with your classmates by looking up the dictionary or surfing the internet.

Look at the following words we may use in this scene.

Place	Replace	Electricity	System
Workshop	Application	Energy	Rotational
Near	Always	Convert	Steam
Nearly	Supply		

Now look at the following Chinese meaning with your colleagues carefully, and match the English and Chinese meaning in a given time, whoever finishes off it first will be the winner. Then dictate them for each other among your team workers.

车间，工场	旋转的，循环的	接近，靠近	将近，大约，几乎
提供	使转变	能力	电，电学
地点，位置	使用，应用	总是，始终	蒸汽
替换，代替	系统，体系，制度		

_____
_____
_____
_____
_____

The following expressions are useful. Before we finish off the task please make out their English meaning with the help of your teammates or dictionary or surfing the internet.

把……变成	把……供……	接近	以……形式
蒸汽机	个别地	个人，个体	依靠

_____
_____

Module Two Mechanical Machining

Energy is required in various forms by machine tools, for example, reciprocating motion to a planer table, longitudinal feed to a lathe carriage, rotational motion to a drill, etc. It is always supplied to the machine in the form of rotational energy at the pulley.

One of the functions of the machine is to convert the rotational energy into the required form. Energy is supplied to most of our workshops in the form of electricity, and it is converted to rotational energy by means of an electric motor.

Work in groups and read through the materials above carefully with your group members and then finish off the following exercises.

Discuss with your colleagues and try to share your ideas and then make sure which of the following expression is right or wrong. T for Right and F for wrong.

(1) Machine tools need various forms of energy.   (   )

(2) We always get the energy for t the machine in the form of rotational energy at the pulley.
     (   )

(3) Machine can convert the rotational energy into the required form.   (   )

(4) Electric motor can convert electricity to rotational energy.   (   )

What kind of motion the following tool can apply? Work with your group members and then try to tick the item out if it does.

to a planer table	to a lathe carriage	to a drill
reciprocating motion	longitudinal feed	rotational motion
( )	( )	( )

Early machine tools were driven by a reciprocating steam engine through a system of pulleys belts, and clutches. Usually, a single steam engine provides the power for several machine tools in the workshop. Later, single electric motor replaced the steam engine to drive a group of machine tools. In modern workshops, most of the machine tools are provided with individual motor drive. Each machine has its own electric motor which drives through belt, gearing, or by direct coupling.

The following expressions are useful. Before we finish off the task please make out their Chinese meaning with the help of your teammates or dictionary or surfing the internet.

machine tools	reciprocating steam engine	pulleys belts
clutches	steam engine	electric motor
individual motor drive	belt	gearing
coupling	workshop	

83

机械制造类专业英语

Look at the material above carefully and then work with your workmates to solve the following question.

Please discuss with your classmates the steps of driving machines in history with the help of your teacher.

_____
_____
_____
_____

Read the materials above again with your work mates and try to understand the meaning of it. Put the following sentences in right order, asking your partners for help if necessary.

( ) Most machine tools has its own individual motor drive.
( ) A single steam engine drives several machine tools.
( ) Single electric motor took place of the steam engine to drive a group of machine tools.
( ) A reciprocating steam engine drives machine tools by pulleys belts, and clutches.
( ) The electric motor drives its own machine by belt, gearing, or by direct coupling.

Read through the information above again with your workmates and try to fill in the blanks in the following short passage.

_____ machine tools were driven by a reciprocating steam engine through a system of pulleys belts, and clutches. _____, a single steam engine provides the power for several machine tools in the workshop. _____, single electric motor replaced the steam engine to drive a group of machine tools. _____, most of the machine tools are provided with individual motor drive. Each machine has its own electric motor which drives through belt, gearing, or by direct coupling.

Observe your workplace and try to understand the power in workshop, read the materials above again and write down the development of power in workshop.

_____
_____
_____
_____

Group work. The students will work in groups discussing the power in workshop and then exchange their experiences in workplace.

Module Two Mechanical Machining

✎ Practice spoken English.

Mr. Liu: Good afternoon, Mr. Xiong.
Mr. Xiong: Good afternoon, Mr. Liu.
Mr. Liu: I have some trouble with heat treatment. Could you tell me?
Mr. Xiong: What are the problems?
Mr. Liu: Do you know what is the function of heat teatment?
Mr. Xiong: Yes. Heat treatment is a method by which the treater can change the physical properties of a material.
Mr. Liu: How many types of heat treatment are there?
Mr. Xiong: There are main operations in the heat treatment of steel: hardening, tempering and annealing.
Mr. Liu: How is hardening operation done?
Mr. Xiong: The hardening operation consists of heating the steel above its critical tange and the quenching it in a suitable medium such as water, brine, oil, or some other liquid. Having been hardened, the metal must be given a tempering treatment which consists of reheating the hardened steel to a temperature below the critical range, thus producing the required physical properties.
Mr. Liu: And what is critical temperature?
Mr. Xiong: It is the temperature at which a certain change takes place in the physical condition of the steel.
Mr. Liu: What is the purpose of annealing?
Mr. Xiong: It can soften a work-piece and relieve internal stresses produced by machining.
Mr. Liu: Thank you telling me about these.
Mr. Xiong: I'm glad to have been of some service.

📖 New words and expressions we learn (let the students copy the new words and expressions he or she doesn't know in this scene and then dictate them among his/her group members)

Self assessments

Time_____ Class_____

No.	Name	Evaluation Items				Notes
		Attitude	Presentation& Passage	Exercises	Activities	

Teachers' Observation Record and Evaluation Sheet

level \ item	listening	speaking	reading	writing	familiarity of parts
good					
better					
best					
bad					

Module Three
CNC Machining

Scene One CNC Machine

Before you step into this scene, please learn the following words and expressions by heart with your classmates by looking up the dictionary or surfing the internet.

machine control unit (MCU)	机床控制单元
alphanumeric	adj. 文字数学的，字母与数学结构的
manual	adj. 手工的，手动的，人工的
coordinate	n. 坐标
servo	n. 伺服
positioning	n.（坐标）定位
continuous-path tool movements	刀具的连续轨迹运动
incremental positioning	增量定位
closed-loop control	闭环控制
open-loop system	开环系统

A computer numerical control (CNC) machine is an NC machine with the added feature of an on-board computer. The on-board computer is often referred to as the machine control unit or MCU. The MCU usually has an alphanumeric keyboard for direct or manual data input (MDI) of part programs.

Can you explain what CNC/ MCU/ MDI are?

Work with your classmates and try to classify the types of the following CNC machine according to different criteria.

By coordinate definition _____
By servo control systems _____
By positioning systems _____
By machine tool control _____
By power drives _____

▲continuous-path(or contouring)tool movements   ▲incremental positioning
▲closed-loop   ▲right-hand coordinate system   ▲open-loop
▲left-hand coordinate system   ▲hydraulic   ▲pneumatic
▲electric   ▲point-to-point tool movements   ▲absolute positioning

Observe your workplace and try to make out the application of NC/CNC.

NC technology has found many applications, including lathes, and turning centres, milling machines and machining centres, punches, electrical discharge machines(EDM), flame cutters, grinders, and testing and inspection equipment.

Do you know the types of CNC equipment?

Machining centres are the development in CNC technology. Turning centers with increased capacity tool changers are also making strong appearance in modern production shops. These CNC machines are capable of executing many different types of lathe cutting operations simultaneously on rotating part.

In addition to machining centers and turning centres, CNC technology has also been applied to many other types of manufacturing equipment. Among these are wire electrical discharge machines (WEDM) and laser cutting machines.

Observe the following figures carefully with your colleagues and then fill in the blanks with the help of your classmates.

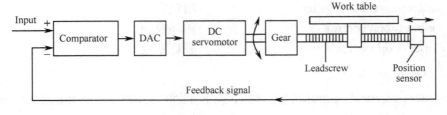

Figure 3-1 _____

Module Three CNC Machining

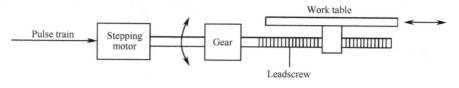

Figure 3-2 _____

( ) A closed-loop system: A closed-loop system utilizes the servo motor whose types include AC servo motorised servo motors.

( ) An open-loop system: An open-loop system utilizes stepping motors to create machine movements.

 Which of the followings belong to the construction of CNC machines? Please tick it out if it does.

☐ control system ☐ drive motors ☐ computer ☐ tool changers

New words and expressions we learn in this session. (let the students copy the new words and expressions he or she doesn't know in this session and then dictate them among his/her group members)

Practice spoken English.

Mr. Zhou: Good morning, Miss Yang.
Miss Yang: Hi, Mr. Zhou.
Mr. Zhou: What are you doing now?
Miss Yang: I am working on surface cut.
Mr. Zhou: What's the definition of surface cut?
Miss Yang: …
Mr. Zhou: Is surface cut important in workshop?
Miss Yang: …
Mr. Zhou: What is the classification of surface cut?
Miss Yang: They are …
Mr. Zhou: Thanks a lot.
Miss Yang: You are welcome.

Grammar 9 复合宾语

复合宾语是指宾语和宾语补足语构成的复合结构。复合宾语中的宾语和宾语补足语，在逻辑上有主语和谓语的关系，因此，有时称这样的宾语和宾语补足语为逻辑主语和逻辑谓语。

复合宾语主要有如下几种类型。

1. 名词或代词的宾语+形容词

He thought the balance of account wrong.
他认为这个账户的余额错了。

The worker painted the wall white.
工人把墙漆成白色。

2. 名词或代词+不定式

Poverty forces many people to sleep in the streets and die of sickness and hunger.
贫困使许多人谁在大街上和死于疾病和饥饿。

注：在有些动词（如 let, make, see, hear watch, find, feel 等）的复合结构中，不定式的 to 需省去，如：

We found people lose interest in saving money.
我们发现人们对存款失去了兴趣。

He made his son write out the exercise again.
他让他的儿子把练习再做一遍。

3. 名词或代词+分词

She heard her baby crying in the bedroom,.
她听见她的孩子在卧室里哭。

We must get the equipment repaired.
我们必须修理这台设备。

4. 名词或代词+名词

We elected them directors at a stockholders meeting.
在一次股东大会上，我们选他们为董事。

The incorporators named the corporation CAP.
公司创办人给公司起名为 CAP。

Scene Two CNC Programming

Can you make clear the meaning of the following figures? Survey your workplace carefully with your team mates and try to match the following descriptions with the fiqures below.

(　　) machine axis for a three-axis vertical CNC machine (Machine axis are defined as spindle movement.)

(　　) machine axis for a three-axis horizontal CNC machine (Machine axis are defined as spindle movement.)

(　　) coordinate system for a horizontal milling machine

(　　) coordinate system for a vertical milling machine

(　　) machine axes for six-axis vertical CNC machine

Module Three CNC Machining

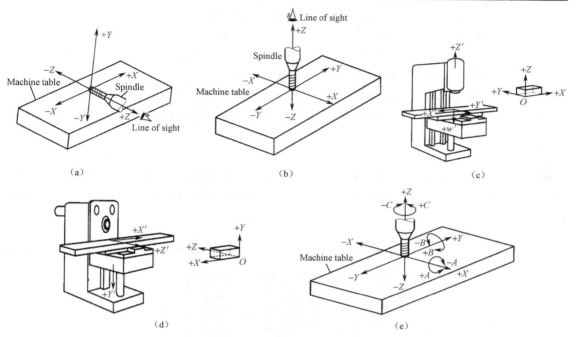

Figure 3-3

Study the NC programme carefully with your teammates and then discuss it with your classmates, lastly read the following materials with your workmates then fill in the blanks with the corresponding answers.

A. Preparatory functions   B. Miscellaneous control   C. Machining parameters
D. Coordinates   E. Tool control   F. F-code
G. M-code   H. T-code   I. R-code
J. S-code   K. G-code   L. Cycle functions
M. Coolant control   N. Interpolators

(   ) the words specify which units, which interpolator, absolute or incremental programming, which circular interpolation plane, cutter compensation, and so on.
(   ) define three dimensional(and three rotational) axes.
(   ) specify feed and speed.
(   ) specifies tool diameter, next tool number, tool change, and so on.
(   ) specify drill cycle, born cycle, mill cycle, and clearance plane.
(   ) specify the coolant condition, that is, coolant on/off, flood, and mist.
(   ) specifies all other control, and spindle on/off, spindle rotation direction, pallet change, clamps control, and so on.
(   ) linear, circular interpolation, circle center, and so on. These control functions are programmed trough program words (codes).

(　　) the G-code is also called preparatory code or word.
(　　) The F-code specifies the feed of the tool motion.
(　　) the S-code is the cutting speed code.
(　　) The T-code is used to specify the tool number.
(　　) The R-code is used for cycle parameter.
(　　) The M-code is called the miscellaneous word and is used to control miscellaneous functions of the machine.

Watch your workplace intensively with your group members and then look at the following picture carefully with your workmates, then discuss the coordinate system for a lathe with your classmates.

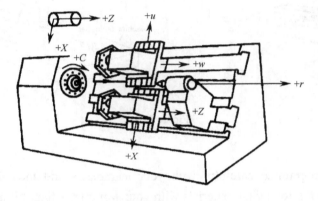

Figure 3-4

The setup operations can begin after the fixtures; tooling, program tape, setup sheets, and part blank arrive at the CNC machine.

Watch, learn and read the following materials carefully with your classmates, and finding the blank.

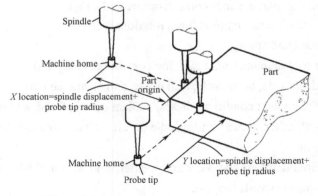

Figure 3-5 _____

Module Three CNC Machining

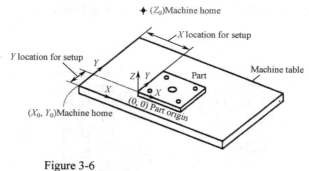

Figure 3-6 _____

Do you know what kind of control systems they are?

(   ) Point-to-point tool movements cause the tool to a point on the part and execute an operation at that point only.

(   ) Continuous-path tool movements are so named because they cause the tool to maintain continuous contact with the part as the tool cuts a contour shape.

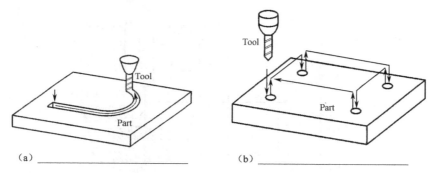

(a) _____   (b) _____

Figure 3-7

Look at the following figure and discuss it with your classmates then try to make clear of the meaning of the terms.

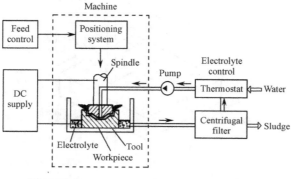

Figure 3-8 _____

Can you give a topic to the following figure with the help of your workmates?

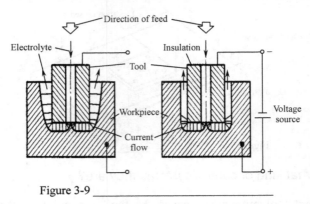

Figure 3-9 _____

Can you describe the principle of an EDM die-sinking machine after observing the following figure?

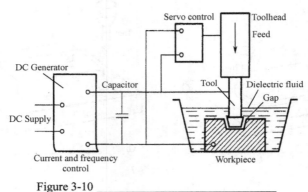

Figure 3-10 _____

Look at the following figure and talk about it with your group members.

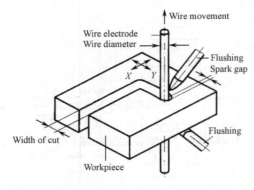

Figure 3-11 _____

Module Three CNC Machining

 New words and expressions we learn in this session. (let the students copy the new words and expressions he or she doesn't know in this session and then dictate them among his/her group members)

Practice spoken English.

Mr. Wang: Good morning, Mr. Yang.
Mr. Yang: Good morning, Mr. Wang.
Mr. Wang: It seems you are very busy now.
Mr. Yang: Yes, a little.
Mr. Wang: What are you doing?
Mr. Yang: I am thinking about chatter phenomenon.
Mr. Wang: Is chatter phenomenon common?
Mr. Yang: …
Mr. Wang: How can we deal with it?
Mr. Yang: …
Mr. Wang: Why does it happen?
Mr. Yang: Because it is virtually impossible to manufacture a part without chatter and we must consider the …
Mr. Wang: Oh, I see. I am very glad to know this. Thanks.
Mr. Yang: You're welcome.

Grammar 10　广告的翻译

广告是较为独立的一种文本，我们需要熟悉这种文本并在翻译中体现其风格和特点。

1. 突出原广告的重点

例：

FINALLY Dbase you have been waiting IV.
Get the new Dbase IV. Now for just $449.

这是一份计算机软件广告，通过变换字体可使人一目了然其重点所在。

译文：BASE IV 终于问世，人们盼望已久的 BASE 新版，目前售价仅 449 美元。

2. 注意广告中的修辞手段

例：

Are you getting fat or lazy waiting for your plotter?
Don't wait——Get PLUMP.

这是一则推销 PLUMP 电脑的广告，采用了极为夸张的语言。

译文：用绘图仪速度太慢，使人等得又懒又胖，不如去买台 PLUMP。

3．反映原广告的通俗性

不少广告使用口语，俚语，俗语或谚语以及其他形式文体，翻译文尽可能与其贴近。

例：

Here's proof that something small can be powerful.

这是一则危机广告。

译文：别看体积小，功能不得了。或：外形虽小，功能惊人。

A whole year without a single bug.

这是一个电子公司的产品广告。

译文：全年无故障。或：万无一失。

4．体现原广告的神韵

许多广告语言生动、传神、深刻、耐人寻味，译文要尽可能不失原味。

例：

I feel like a donkey. For not buying the "Access" portable computer.

译文：不买"Access"便携式计算机使我劳累得像头笨驴。

5．用汉语 4 字词组翻译常用广告语

有许多产品广告语看似简单，要转化为相应汉语却颇费周折，如果采用汉语中常用的 4 字词组，难题往往迎刃而解。

例：

a wide selection of styles	品种多样
beautiful range of colours	花式齐全
bright colours	色彩鲜艳
in up-to-date style	款式入时
compact	结构合理

以上翻译要点均指广告的主题部分，其实一份完整的产品广告一般都有 4 个内容：标题；产品图；正文（常与图配合）和其他（商标、名称、经销地、地址等）。

Scene Three　　CNC Machining Center

　　Before you step into this scene, please learn the following words and expressions by heart with your classmates by looking up the dictionary or surfing the internet.

　　Look at the following words we may use in this scene and try to make out their Chinese meaning in a given time, whoever finishes it off first will be the winner.

original	random	potential
machining center	holding fixture	

Module Three CNC Machining

Now look at the following Chinese meaning with your colleagues carefully and try your best to make out their English meaning in pairs, finally dictate them for each other among your team workers.

adj. 任意的，随便的   adj. 最初的，原始的   n. 夹具
n. 加工中心   n. 潜能，潜力

The flexibility and versatility of numerical control have led to the development of a new type of machine tool called the machining center, which can use simpler work holding fixtures and fewer cutting tools.

We provide an automatic tool changer for the machining center. The tool change arm located above the spindle rotates clockwise, simultaneously gripping the tool in the spindle and another tool in an interchange station located on the face of the machine which is used to store the tool temporarily. The arm then moves forward, removing the tools from the spindle and from the interchange station. After rotating clockwise 180 degrees, the arm retracts, inserts the new tool in the spindle, and places the used tool in the interchange system. The arm then returns to the original position. A mechanical hand inside the drum removes the tool from the interchange station and stores it in the tool drum. The mechanical hand removes the next tool to be used from the drum and stores it in the tool drum. The mechanical hand removes the next operation. The tool change operation may be completed in 5 seconds. The tool drum can hold a large number of different tools. Each tool holder being coded, the tools can be selected in a random order and in any sequence.

The potential of these machines can be fully realized with the help of NC.

Read the underlined materials above again with your workmates and then try to translate them into Chinese, and finally discuss with your workmates.

（1）We provide an automatic tool changer for the machining center.

_____

（2）The arm then moves forward, removing the tools from the spindle and from the interchange station.

_____

（3）The arm retracts, inserts the new tool in the spindle, and places the used tool in the interchange system.

_____

（4）A mechanical hand inside the drum removes the tool from the interchange station and stores it in the tool drum.

_____

（5）Each tool holder being coded, the tools can be selected in a random order and in any sequence.

___

Read the materials above first with your colleagues and then try your best to write down the definition of machine center.

___

Work with your group members and read the information above again, then describe the process of tool change.

___

Read the paragraph of the materials above and do your best to answer the following questions with the help of your classmates.

（1）What does the tool change arm first do in the spindle and in an interchange station?

___

（2）What is located on the face of the machine and is used to store the tool temporarily?

___

（3）What does the arm do next from the spindle and from the interchange station?

___

（4）Can the tools be selected in a random order and in any sequence?

___

Work with your group members and read the above information again and try your best to translate the following terms into Chinese.

in a random order	tool holder	tool drum
NC	mechanical hand	machining center
automatic tool changer	in the spindle	in any sequence
rotating clockwise	moves forward	new tool
used tool	interchange station	

___

Survey your workplace and observe carefully, make out the process of tool change and ask the master for help if there are any problems, write down your experiences in your workplace.

___

Survey your workplace with your team mates and watch the operation panel of CNC machines carefully and then finish off the following tasks.

Module Three CNC Machining

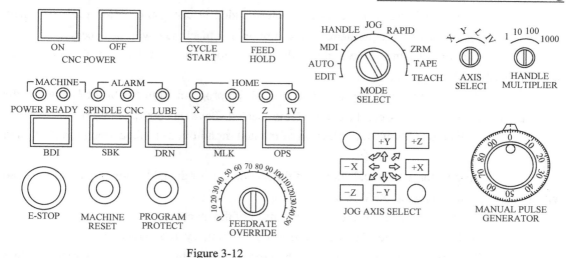

Figure 3-12

① CONTROL PANEL
② POWER ON/OFF
③ RESET
④ MDI
⑤ HANDLE
⑥ JOG
⑬ CRT CHARACTER DISPLAY
⑭ DISPLAY
⑮ EMERGENCY STOP
⑯ MODE SELECT
⑦ CYCLE START
⑧ RAPID TRAVERSE OVERRIDE
⑨ MACHINE LOCK
⑩ FEED HOLD
⑪ SINGLE BLOCK
⑫ DRY RUN
⑰ EDIT
⑱ MEMORY
⑲ REFERENCE RETUR
⑳ FEED RATE OVERRIDE

(   ) The contents of the currently active program are active sequence number display.

(   ) The screen is on which all the characters (addresses) and data are shown.

(   ) This button is used for alarming cancellation or cancellation of an operation.

(   ) The control is always turned on after the main power switch, located on the door of the control system.

(   ) Usually, the control panel is located at the front of the main electronic system and is equipped with a screen and with various buttons and switches.

(   ) This switch is used to check a new program on the machine. All movements of the tool are blocked, while a check is run on the computer screen.

(   ) This switch reduces the rapid feed rate.

(   ) This allows the control of the work feeds defined by function F. You can increase or decrease the percentage of the value entered in the program.

(   ) Turning this switch ON and then pressing button X, Y or Z causes the machine to return to the zero position for each axis.

(   ) By turning this switch ON, all the rapid and work feeds are changed to one.

( ) This switch can also be used if you intended to check the performance of a new program on the machine or if the momentary interruption of a machine's work is necessary.

( ) The control lamp located above the button goes on, and the control lamp located above CTCLE START goes off.

( ) Use this button during the execution of the program from MEMORY or when the CYCLE START button is pressed, the control lamp located above this button goes on.

( ) This mode allows the selection of manual monotonous to feed along the X, Y, and Z axes.

( ) Allows for the manual control of the movements of the table with the use of the hand wheel, along one of four axes; X, Y, Z or B.

( ) This mode enables the automatic control of the machine.

( ) This mode enables the NC to command the memory to be executed.

( ) This mode enables you to: Enter the program to machine memory, Enter any changes to the program, Transfer data from the program to a tape, Check the memory storage capacity.

( ) Use this switch to specify the operational mod.

( ) This button is red and is located in the lower left corner of the control panel.

There are two main types of maching centres: the horizontal spindle and the vertical spindle machine. Please give out the names for the following figures.

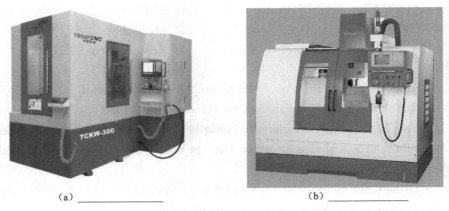

(a) _____   (b) _____

Figure 3-13 _____

Learn the following materials with your colleagues and then do your best to fill in the blanks with the help of your partners.

In a CNC system, each _____ of motion is equipped with a separate driving source that replace the hand wheel of the conventional machine. The driving source can be _____ motor, a _____, or a _____. The source selected is _____ mainly based on the precision requirements of the machine.

Module Three CNC Machining

Survey your workplace with your classmates carefully and then try to fill in the blanks in the following figure with the help of your partners, finally do your best to memorize the terms in the figure.

The main components of CNC machining centres are the bed, saddle, column, table, servomotors, ball screws, spindle, tool changer, and the machine control unit.

Figure 3-14 _____

Observe your workplace with your colleagues the different kinds of machine center and the components of a CNC machine center and then try your best to finish off the following tasks: task 1, fill in the blank in the following figure; task 2, remember the terms with the help of your group members, then dictate them for each other.

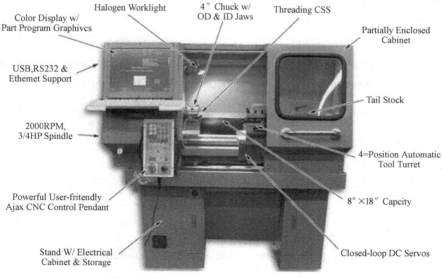

Figure 3-15 _____

Practice spoken English.

Mr. Zhang: Hello, Mr. Cheng.

Mr. Cheng: Hi, Mr. Zhang.

Mr. Zhang: I want to know some knowledge of the CAD. Could you tell me some about it?

Mr. Cheng: No problem. What do you want to know?

Mr. Zhang: What does the CAD mean?

Mr. Cheng: Computer aided design (CAD) means the use of a computer to assist in the design of an individual part or a system. The design process usually involves computer graphics.

Mr. Zhang: What is the CAD system?

Mr. Cheng: It is basically a design tool in which the computer is used to analyze aspects of a designed product.

Mr.Zhang: What can the CAD system support?

Mr.Cheng: It can support the design process at all levels, conceptual, preliminary, and final design.

Mr. Zhang: Can the designer test the product by the CAD system?

Mr. Cheng: Yes. It can do.

Mr. Zhang: What is one of the most valuable features of CAD systems?

Mr. Cheng: The display of the designed object on a screen.

Mr. Zhang: What systems are the most CAD systems using?

Mr. Cheng: Interactive graphics systems.

Mr. Zhang: Do you think the elimination of hidden lines is one of the most difficult problems in CAD drawings?

Mr. Cheng: Yes, I think so.

Mr. Zhang: Oh. I have learned so much from you. Thanks a million.

Mr. Cheng: At your service.

New words and expressions we learn（let the students copy the new words and expressions he or she doesn't know in this scene and then dictate them among his/her group members）.

Grammar 11　动名词

动名词也是非限定动词的一种，它是由动词原形+ing构成，它有名词和动词的特征。以下为它的用法。

1．动名词作主语

Interpreting figures is an important part of the accountant's function.
解释数据是会计职能的一个重要部分。

Determining the least costly way to produce a specific quantity of output occupies much of the time of managers.
设法以最低成本生产定量产品要耗费管理者大量时间。

2．动名词作表语

The bookkeeper's work is recording financial data.
簿记员的工作是记录财务数据。
His interest is reading novels.
他的兴趣是读小说。

3．作宾语

直接宾语，可以用在 begin，start，stop，finish，like 等动词后面。
They haven't finished packing the goods.
他们还未完成货物的包装。
He likes playing basketball.
他爱打篮球。

4．作介词宾语

We took part in installing the equipment.
我们参加了安装这些设备。
He is responsible for organizing the meeting.
他负责组织这个会议。

Scene Four　CAD/CAM Software

Read the following materials about CAD/CAM software with your classmates and try to understand the meaning of the materials. Finish the exercises following the materials.

Many tool paths are simply too difficult and expensive to program manually. For these situations, we need the help of a computer to write an NC part program.

The fundamental concept of CAD/CAM is that we can use a CAD drafting system to draw the geometry of a workpiece on a computer. Once the geometry is completed, then we can use a CAM system to generate an NC tool path based on the CAD geometry.

What's CAD/CAM software? Please try your best to answer the question with the help of your colleagues.

Work in your workplace and observe the CAD/CAM system carefully and then try to fill in the blanks in the following figures.

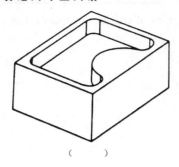  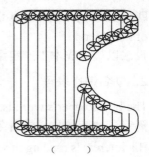

(　　)　　　　　　　(　　)　　　　　　　(　　)

Figure 3-16

A. the model is verified
B. the CAM system creates tool paths
C. the geometry is defined in the CAD drawing

Watch the CAD drawing carefully in your workplace and discuss with your group members the four steps from a CAD drawing to the working NC code then put the following sentences in right order with the help of your team mates.

(　　) The final product of CAD/CAM process is the NC code.
(　　) The geometry is defined in a CAD drawing.
(　　) The model is next imported into the CAM module.
(　　) The CAM model is then verified to ensure that the tool paths are correct.

New words and expressions we learn in this session. (let the students copy the new words and expressions he or she doesn't know in this session and then dictate them among his/her group members).

Grammar 12　否定的处理

同每一种语言一样，英语也有自己的否定表达方式。由于使用的方式及词汇、语法手段与汉语有所差异，所以翻译时应灵活地从正反两方面进行处理。

1. 完全否定

英语中表示完全否定的词有 NO（没有），NONE（没有人、无一物、一点也不），NOT（不），NEVER（从来不，绝不），NEITHER（两者都不），NOR（也不）和 NO 与其他词组成的合成词等。由上述词构成的否定句一般仍为汉语否定句，可直译。

例如：

No defect did we find in these circuits.
在这些电路中，我们没有发现任何缺陷。

Videophone service will spread from the office to the home in the not far distance future.
在不久的将来可视电话业务将会从办公室扩展到家庭。

Nothing in the world moves faster than light.
世界上没有任何东西可比光速更快。

2. 部分否定

英语中的部分否定 通常是由 not 与 all both every always many 等词连用构成。通常译为"不都是、不全是、并非、不总是、不多"，等等。

例如：

All that glitters is not gold.
发光的不都是金子。

Every amplifier can not be dependable.
并非每个放大器都可靠。

I do not have much experience in operating microcomputer.
我并没有多少操作微机的经验。

3. 双重否定

在句子中出现两次否定，主要是为了强调，翻译时应变为肯定句。常见的词组有：no…not +…（no）（没有…没有），without…nor…（没有…就不…），not…until…（直到…才…），not…but…（没有…不…），never…without…（每逢…总是…），no…other than…（不是…正是…），等等。

例如：

There is no grammatical rule that does not have exceptions.
做法规则都有例外。

You cannot be too careful in performing an experiment.
做实验越仔细越好。

Heat can never be converted into certain energy without something lost.
热能转换成某种能量总是有些损耗。

4. 意义否定

英语中很多含有否定意义的词和词组，一般可翻译成否定。这些词和词组可以是动词、名词、形容词、副词、连词和介词等。

例如：

The operator must have missed the noticed and instruction on the instrument.
操作人员一点没有看到仪器上的注意事项和需知。

One of the advantages of optical fibers is freedom from direct effect of interfering radio transmission.
光纤的优点之一就是不受无线电传输干扰的直接影响。

Our knowledge of propagation phenomena at optical wavelength is from perfect.
我们对光波传播现象的认识还很不全面。

It's hardly a possibility to mention all the kinds of applications of electronic computers.
将电子计算机的应用种类列举齐全几乎是不可能的。

Unless there is motion, there is no work.
没有运动就没有做功。

But for air and water, nothing could live.
没有空气和水，什么也不能活。

Scene Five CAD/CAM Components and Functions

 1 <u>CAD/CAM systems contain both CAD and CAM capabilities, each of which has a number of functional elements.</u> The CAD portion of the system is used to create the geometry as a CAD model. The CAD system, whether stand alone or as part of a CAD/CAM package, tends to be available in several different levels of sophistication. Wire frames can be difficult to visualize, but all Z-axis information is available for the CAM operations. 2 <u>Solid models can be sliced open to reveal internal features and not just a thin skin.</u>

Please talk about the material above with your teammates and try to understand the meaning of it, and then translate the underlined sentences into Chinese with the help of your workmates.

_____
_____

Work with your classmates observing the following figures carefully and discuss them with your group members in details then try to give a topic to each of the following pictures, finally put them in right order with the help of your classmates.

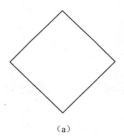 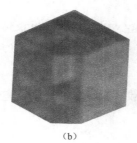  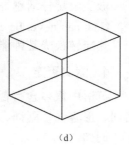

(a)　　　　　　　　(b)　　　　　　　　(c)　　　　　　　　(d)

Figure 3-17

Find out the correct Chinese translation for each of the 3 English sentences.

A. the CAM module is used to create the machining process model based upon the geometry supplied in the CAD model.

B. the CAM system will generate a generic intermediate code that describes the machining operation, which can later be used to produce G and M code or conversational programs.

C. the CAM modules also come in several classes and levels of sophistication.

Module Three CNC Machining

（　）CAM 模块也有多种类型和层次。

（　）CAM 系统生成描述加工操作的普通中间代码，这些代码以后可以用来生成 G 和 M 代码或会话式程序。

（　）根据 CAD 模型提供的几何形状，CAM 模块用于创建加工工艺模型。

We know a standard CNC machine tool can't be used directly to produce an ellipse. How do we solve this problem?

Read the following materials and observe the figure followed and then try to find out the answers.

The CAM system will reconcile this problem by estimating the curve with line segments.

The CAM system will generate a bounding geometry on either side of the true curve to form a tolerance zone.

Smaller tolerance zones will produce finer tool paths and more numerous line segments, while larger tolerance zones will produce fewer line segments and coarser tool paths.

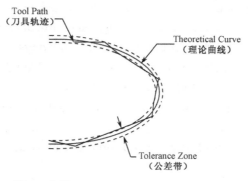

Figure 3-18 _____

Discuss with your teammates about the following sentences.　　　　True　False

（1）One important concept we must understand is that the geometry represented by the CAD drawing may not be exactly the same geometry that is produced on the CNC machine tool.
　　　　　　　　　　　　　　　　　　　　　　　　　　　　　　　□　□

（2）To create the machining operations, the CAM system will need to know which cutting tools are available and what material we are machining.
　　　　　　　　　　　　　　　　　　　　　　　　　　　　　　　□　□

（3）It is easy to rely on the computer to generate the correct tool path, but finished surfaces are further estimated during machining with ball end mills.
　　　　　　　　　　　　　　　　　　　　　　　　　　　　　　　□　□

（4）CNC machine tools are equipped to produce very accurate tool paths as long as the tool paths are either straight lines or circular arcs.
　　　　　　　　　　　　　　　　　　　　　　　　　　　　　　　□　□

（5）NURBS curves can represent virtually any geometry, ranging from a straight line or circular arc to complex surfaces.
　　　　　　　　　　　　　　　　　　　　　　　　　　　　　　　□　□

（6）More progressive CAM developers tend to produce their tool and material libraries as database files that can be easily modified and customized for other applications.

Observe your workplace carefully with your colleagues and then read the following materials trying to match them with right Chinese translations.

(　　) This can be via a simple back plot of the tool centreline or via a sophisticated solid model of the machining operations.

(　　) The post-processor is a software program that takes a generic intermediate code and formats the NC code of each particular machine tool control.

A. 后置处理器是一个软件程序，他将通用的中间代码格式化为使用于每个特定机床控制器的 NC 代码。

B.（这种验证）可以通过在屏幕上画出简单的刀具中心线或复杂的实体模型（来实现）。

Read the following materials and try to know software issues and trends in your workplace. Translate them into Chinese for each other and then discuss them in pairs.

Pure CAD systems are used in all areas of design, and virtually any product today is designed with CAD software. Gone are the days of pencil and paper drawings.

This mismatch sets up the classic argument between the CAD designers and the CAD/CAM programmer on what is the best way to approach CAD/CAM.

The greater the acceptance of the standard, the greater the return on investment for the businesses that own the software.

Regardless of the path that is chosen, organizations and individuals tend to become entrenched in particular technology.

Table 3-1　G Code Reference

G codes	Definition and Syntax	定义和语法
G00	Rapid traverse G00 X Y Z	快速定位 G00 X Y Z
G01	Linear interpolation G01 X Y Z F	线性插补 G01 X Y Z F
G02	Circular interpolation(clockwise) G02 Z Y Z I J (or R) F	圆弧插补（顺时针） G02 X Y Z I J (or R) F
G03	Circular interpolation (counterclockwise) G03 X Y Z I J (or R) F	圆弧插补（逆时针） G03 X Y Z I J (or R) F
G04	Programmed dwell	编程暂停
G17	X-Y are plane selection	X-Y 圆弧平面选择
G18	X-Z are plane selection	X-Z 圆弧平面选择
G19	Z-Y are plane selection	Z-Y 圆弧平面选择
G20	Inch programming units	编程单位为英寸

Module Three CNC Machining

续表

G codes	Definition and Syntax	定义和语法
G21	Millimeter programming units	编程单位为毫米
G28	Return to machine zero through intermediate point (G91) G28 Z Y Z	从中间点返回机床原点（G91）G28 X Y Z
G29	Return from machine zero to an intermediate point	从机床原点返回中间点
G40	Cancel cutter compensation	取消刀具补偿
G41	Cutter compensation left	刀具左侧补偿
G42	Cutter compensation right	刀具右侧补偿
G43	Select tool height offset	选择刀具高度偏置
G49	Cancel tool height offset	选择刀具高度偏置
G54-59	Workplane selection	工作平面选择
G73	Drill cycle with chip breaker	切屑钻孔循环
G74	Tapping cycle (LH)	攻螺纹循环（LH）
G76	Boring cycle with orient and rapid retract	带主轴准停和快速回退的镗削循环
G80	Canned cycle cancel	取消固定循环
G81	Drilling cycle G81 X Y Z F R L	钻削循环 G81 X Y Z F R L
G82	Drilling cycle with dwell G82 X Y Z F R L P	带中断的钻削循环 G82 X Y Z F R L P
G83	Peck drilling cycle G83 Z Y Z F Q R L	步进式钻削循环 G83 X Y Z F Q R L
G84	Tapping cycle G84 X Y Z F R L	攻螺纹循环 G84 X Y Z F R L
G85	Boring cycle G85 X Y Z F R L	镗孔循环 G85 X Y Z F R L
G86	Boring cycle (rapid retract)	镗孔循环（快速退刀）
G87	Boring cycle (manual retract)	镗孔循环（手动退刀）
G88	Boring cycle (manual retract with dwell)	镗孔循环（带中断的手动退刀）
G89	Standard boring cycle with dwell	带中断的标准镗孔循环
G90	Absolute coordinate system	绝对坐标系
G91	Incremental coordinate system	增量坐标系
G92	Shift work coordinate system	转换工作坐标系
G98	Return tool to initial plane after canned cycle	刀具在固定循环之后返回初始平面
G99	Return tool to retract plane after canned cycle	刀具在固定循环之后返回退刀平面

Table 3-2　M Code Reference

m codes	Definition and Syntax	定义和语法
M00	Unconditional program stop	程序无条件终止
M01	Optional program stop	程序选择性终止
M02	Program end	程序结束
M03	Spindle on clockwise	主轴顺时针旋转

续表

m codes	Definition and Syntax	定义和语法
M04	Spindle on counterclockwise	主轴逆时针旋转
M05	Spindle off	关闭主轴
M06	Automatic tool change	自动换刀
M07	Coolant on (mist)	打开冷却液（雾状）
M08	Coolant on [flood (default)]	打开冷却液［水流状（默认）］
M09	Coolant off	关闭冷却液
M19	Oriented spindle stop	主轴定向停止
M30	Program stop and reset	程序终止并重置
M97	Execute an local subprogram	执行本地子程序
M98	Execute an external subprogram	执行外部子程序
M99	Return control to main program after subprogram or subroutine	执行子程序或子例程之后返回主程序

Table 3-3　Letters

m codes	Definition and Syntax	定义和语法
A	Angular coordinate around X-axis	绕 X 轴的角坐标
B	Angular coordinate around Y-axis	绕 Y 轴的角坐标
C	Angular coordinate around Z-axis	绕 Z 轴的角坐标
D	Tool diameter offset designation	指定刀具直径偏置
F	Linear feed rate	线性进给率
G	Preparatory code	准备代码（G 代码）

New words and expressions we learn in this session.（let the students copy the new words and expressions he or she doesn't know in this session and then dictate them among his/her group members）

Practice spoken English.

Miss Candy: Hello, Mr.Smith Nice to meet you!

Mr. Smith: Hi Miss Candy. Nice to meet you, too.

Miss Candy: What are you doing?

Mr. Smith: I am preparing for some experiments.

Miss Candy: I know your major is engineering. Could you tell me what about the transfer machine?

Mr. Smith: …

Miss Candy: Is transfer machine common in workshop?

Mr. Smith: …

Miss Candy: How transfer machine works?

Module Three CNC Machining

Mr. Smith: …
Miss Candy: I'm really very grateful to you.
Mr. Smith: Not at all.

Grammar 13　长句的翻译

科技英语文章要求合乎逻辑，句子结构严谨。长句按其结构，大部分属于主从复合句和并列复合句，也有一些简单句。

翻译长句，首先要把全文看上几遍，弄懂大意。然后再把长句看上几遍，设法抓住其主要内容。如果这时仍然感到无从下笔进行翻译，可对句子加以分析，弄清长句各部分之间的语法关系、层次关系和前后关系，确定中心内容后根据汉语的习惯表达方式加以转换。

To illustrate, the solution of the mathematical equations for the trajectory of a space rocket would require nearly two years of work by one computer operating an ordinary adding machine, while an electronic computing machine can do the same job in seconds without any error.

该句可以分为3部分。

主句部分：…the solution of the mathematical equations for the trajectory of a space rocket would require nearly two of work by one computer operating an ordinary adding machine。

并列句部分：…while an electronic computing machine can do the same job in seconds without any error。

插入语部分：To illustrate。

然后得出译文：举例来说，一个计算工作者用一架普通加法机来解决一宇宙火箭轨迹的数学方程式，差不多需要两年的工作时间，而一台电子计算机却可以毫无误差地在几秒钟之内完成同样工作。

The unit of p.d. like that of e.m.f. is the volt which is that p.d. existing between two points in a circuit where one unit of electrical energy has been changed to some other form of energy as a consequence of one coulomb passing between the points.

该句为复合句，也可分为3部分。

主句部分：The unit of p.d. like that of e.m.f. is the volt…

定语从句部分：which is that p.d. existing between two points in a circuit…

以上修饰主句中的 volt。

另一定语从句部分：where one unit of electrical energy has been changed to some other form of energy as a consequence of one coulomb passing between the points。

用以修饰 which 定语从句中的 circuit。

全句译为：电位差的单位同电动势的单位一样是伏特。伏特就是存在于电路两端之间的电位差。在这一电路上由于每一库伦的电荷通过两端，电能的每个单位转换成了能的其他形式。

注意，此句译为三句，主句和两个定语从句分别译为一句。

由于英语表达方式尤其在顺序上与汉语不尽相同，在翻长句时为了调整这种顺序，常采用3种不同的方法。

1. 顺译

按英语原句表达顺序用汉语依次译出，其中，"主+谓+宾"结构的英语长句多采用此法。

In addition to malleability and ductility, some metals possess the power to conduct electric current and the ability to be magnetized, two special properties not possessed by any other class of materials.

除展性和延性外，某些金属还具有导电能力和磁化能力，这是任何其他材料所不具备的两种特性。

Lecture "Holography" was delivered by D. Gabor on the occasion of his receiving the 1971 Noble Prize for physics for his invention and development of the holographic method.

D. 盖伯由于发明和发展了全息照相术而获得1971年诺贝尔物理学奖。本文是他授奖时宣读的报告《全息照相》。

2. 倒译

倒译即改变英语原句的顺序，又称变序译法，许多长句用倒译可以比较顺利地译出，顺译反倒难度很大甚至不可能准确译出。

In the case where the waves are incident on a boundary separating two regions where their velocity is different they will in general be divided into a reflected train and a transmitted train whose relative intensities will depend on the magnitude of the velocity change at the boundary, on the abruptness of this change, and on the angle of incidence.

在波射到速度不同的两个地区的分界面上的情况下，它们一般都会放射波列和透明波射列，这两者的相对强度取决于界面上速度变化的大小，变化的缓急度以及入射角的大小。

In this way the distinction between heavy current electrical engineering and light current electrical engineering can be said to have disappeared, but we still have the conceptual difference in that in power engineering the primary concern is to transport energy between distant points in space; while with communications system the primary objective is to convey, extract and process information, in which process considerable amounts of power may be consumed.

在这一方面，强电工程和弱电工程间放入区别可以说已经消失；但我们认为它们在概念上有所不同，因为电力工程的主要任务是在空间相距较远的各地之间输送能量，而通信系统的主要目的是传递、提取和处理信息，尽管在这个时候中或许消耗相当大的电力。

3. 混合译

实际上完全倒译或完全顺译的句子并不多，而两者同时使用的居多。

The two-electrode tube consists of a tungsten filament, which gives off electrons when it is heated, and a plate toward which the electrons migrate when the field is in the right direction.

二极管由一根钨丝和极板组成，钨丝受热时就放出电子，当电场方向为正时，电子就移向极板。

Less voltage is required to breakdown a given thickness air, and more suitable in this respect for a high voltage condenser.

击穿一定厚度的空气比击穿同等厚度的玻璃所需的电压要小，所以玻璃比空气绝缘性

Module Three CNC Machining

好，而且在这一个方面对高压电容来说玻璃较为合适。

Exercises

How about CAD/CAM hardware and software?

Can you talk about the applications of CAD/CAM in the design?

Do you know the heart of them?

Work in groups and survey the workplace, try to figure out how the CAD/CAM system applied in practice.

List three main areas of computer-aided manufacture.

Fill in the blanks in the following passage then talk about it.

Operating systems are programs _____ for a specific computer or class of computers. For convenient and efficient operation, _____ and data are available in the system's memory. The operating _____ is especially concerned with the input/output (I/O) devices like _____, printers, and tape punches. In most _____ the operating system is supplied with the computer.

Translate the following expressions into English.

二维线条图	三维线框模型	刀具轨迹
三维表面模型	三维实体模型	理论曲线
公差带	格式	可定制的

Self assessments

Time _____ Class _____

No.	Name	Evaluation Items				Notes
		Attitude	Presentation & Passage	Exercises	Activities	

113

Teachers' Observation Record and Evaluation Sheet

level \ item	Listening	speaking	reading	writing	familiarity of parts
good					
better					
best					
bad					

Scene Six   FANUC Manual Guide

Operation Guidance function
FANUC MANUAL GUIDE *i*

1. What is MANUAL GUIDE?

❏ Operation Guidance, which supports whole operations on an all-one screen for daily machining including creating a program on a lathe, machining center and compound machine

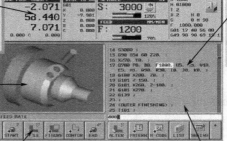

All-in-one Screen
Only one screen concentrated all operations

Machine status window
Machine status such as actual position, feedrate and load meter are displayed always

Easy programming
Based on ISO-code program format, complex muachining motions can be created easily by menu form

Realistic machining simulation
3-D solid model machining simulation is available

CAD/CAM

Intuitive menu selecting
Menu can be selected easily and intuitively by soft-key with icon

Good affinity with CAD/CAM
Most popular ISO-code program format on CAD/CAM can be dealt as it is

2. Market trend of Conver sational programming

❏ Thanks to reducing a price, CAD/CAM on a PC are getting more popular

❏ Desire for more easy operations of a machine and CNC is getting strong

Module Three CNC Machining

☐ The demand of MANUAL GUIDE, which has necessary and sufficient features and thoroughly simplified operations, is getting stronger rather than orthodox conversational programming functions such as Super CAPi T/M and Symbol CAPi T.

FANUC's conversational programming functions will be concentrated to MANUAL GUIDE

Simple operation on all-in-one screen
(10.4"Color / 9.5" Monochrome LCD)

All-in-one screen for Programming, Graphic simulation and Machining

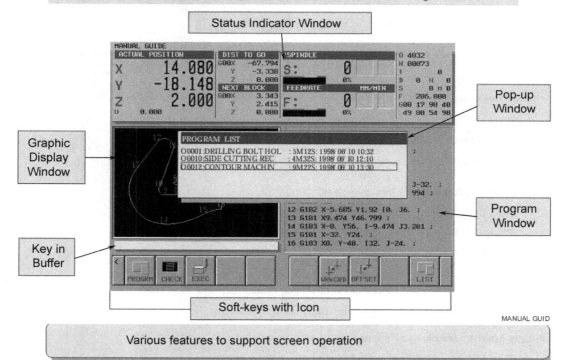

Various features to support screen operation

1. Icon menu

All menu is displayed with Icon soft-keys

Operator can select machining type easily by intuition

(Cycle machining menu soft-keys)

115

2.Application of Pop-up window

All data is displayed on one screen without screen switching

Supplemental data which can not displayed on one screen is displayed an pop-up window

 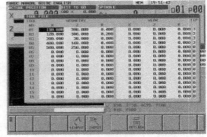

(Tool file window)

FANUC MANUAL GUIDE *i* for Milling

Overview

1.Simple operation on all-in-one screen(10.4"color/9.5"monochrome LCD)
   All operation can be done on one screen,and no screen swithching is needed.

2.Guided ISO programming by conversationalmethod
   (ISO program has higher flexbility than machining process program.)
   a)Easy operations using Icon menu soft-keys
   b)Guidance window to illustrate required parameter

3.Advanced canned cycles for complicated milling machining
   Drilling,Facing,Side cutting,Pocketing,Contour machining

4.Easy program checking by sophisticated graphic simulation
   Realistic animated drawing simulation with solid model

5.Machining used handwheel for manual operation
   Guidance cutting and teach-in for playback

6.Optional function on Series 16*i*/18*i*/21*i*-MA/MB

Module Three CNC Machining

Various features to support milling machine operations

(Data input screen)

(Animated drawing screen)

1. All operations can be done on one screen
2. Advance canned cycles for milling machining
   - a) Drilling
   - b) Hole pattern
   - c) Facing
   - d) Side cutting
   - e) Pocketing
   - f) Contour machining
3. Quick and realistic machining simulation
   Tool path drawing, animated drawing on a solid model, rotation of a product and so on are available
4. Abundant customizing tools for building up the best suited milling machine system
   Installing MTB's own tool set-up guidance, machining process and so on are available
5. Abundant displaying language
   In addition to standard 6 languages display, Japanese / English / German / French / Italian / Spanish, 4 extra languages can be installed easily

Easy program checking

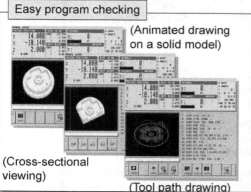

(Animated drawing on a solid model)
(Cross-sectional viewing)
(Tool path drawing)

Contour programming

Automatic calculation of intersections enables simple programming of complicated machining profile

Contour editing function
- Symmetry
- Translation
- Rotation

(Contour programming screen)

Advanced canned cycles

- Contour machining (Side / Pocket / Groove)
  By entering contour figure, tool path of contour machining will be created automatically

(Tool path of contour pocketing)

- Drilling
  (Center drill / Drill / Tap / Reamer Bore / Back bore)
- Hole pattern
  (Line / Arc / Circle / Square / Grid)
- Facing
  (Square / Circle / Ring)
- Side cutting
  (Square / Circle / Track / One side)
- Pocketing
  (Square / Circle / Track / Groove)

Guidance cutting to support manual operation

Guidance cutting, with teaching of auxiliary functions and others, support manual operations strongly

1. Manual operations such as positioning and cutting
2. Auxiliary functions, such as coolant on/off and spindle rotation
3. Approach cutting toward Line and Circle by a Guidance Handwheel
4. Along cutting of Line and Circle by a Guidance Handwheel
5. These motions can be teach-in for playback

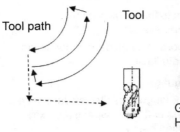

Tool path   Tool

Guidance Handwheel

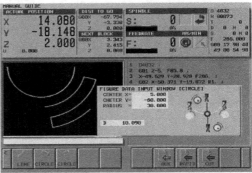

(Sample of guidance cutting screen)

Tool management function

Aoolicable CNC:FANUC Series 16i/18i/21i-MB

1. What is Tool Management function?
   - Tool Management Data Table is available, into which all tool data (Tool Offset value, Tool Life data and so on) are integrated.
   - Tool Management Data Table can be managed collectively by CNC and can be accessed by PMC, MANUAL GUIDE and a personal computer.

2. Adaptation of MANUAL GUIDE to Tool Management function
   - Cutting condition(Feedrate, Spindle Speed) can be set automatically by inputting Tool Type Number.
   - MTB can add their original data to Tool Management Data Table and use it in cycle machining (Available by MTB's customizing).

Module Three CNC Machining

Other new features

(Background Drawing:Example of Full Screen Display)

1. Addition of Background Drawing Function
   - Checking of machining motion can be done easily by machining simulation during other actual machining.

2. Addition of Automatic measuring cycles
   - X/Y/Z single surface   -Stub/Groove
   - Outside/Inside circle   -Outside/Inside rectangula
   - Outside/Inside corner   -Hole position
   - Work-piece angle
   - Automatic calibration for a prove

(Example of an actual measuring motion)

3. Others
   - Full screen display of machining simulation
   - NC Program Expansion for milling canned cycle
   - Hand held calculator type data calculation

FANUC MANUAL GUIDE *i* for Lathe

Overview

1. Simple operation on all-in-one screen(10.4"color/9.5"monochrome LCD)
   All operation can be done on one screen,and no screen switching is needed.
2. Exclusive built operating methods for manual lathe
   a) Easy operations using Icon menu soft-keys
   b) Guidance window to illustrate required parameter
   c) Consistent operations from trial cutting until mass producing
3. Various machining menus
   Bar machining,Grooving,Threading,Lathe drilling,C-aixis drilling/grooving and so on
4. Easy program checking by sophisticated graphic simulation
   Realistic animated drawing simulation with solid model
5. Machining used handwheel for manual operation
   Guidance cutting and teach-in for playback
6. Optional function on Series 16*i* / 16*i* / 21*i*-TA/TB

Various features to support lathe machining operations

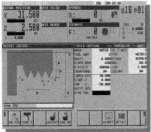

(Programming screen of Threading)

Operation Guidance which possess both of Manual Lathe's Easiness and CNC Lathe's Functionality, and shop-floor programming is also available.

1. Advance canned cycles for lathe machining

    -Bar machining   -Grooving
    -Threading   -Lathe driling
    -C-axis drilling   -C-axis grooving
    -Thread repair cycle

2. All operations can be done on one screen
3. Quick and realistic machining simulation

    Tool path drawing, animated drawing on a solid model, rotation of a product are available

4. Teaching playback

    By teaching operation of skilled worker, its playback machining can be done easily.

(Animated drawing screen)

Various Machining menu for manual operations

1. Guidance Cutting

- Cutting of Line or Circle form can be done by Guidance Handwheel.
- Teaching, Editing of taught-in blocks and Playback machining are available.

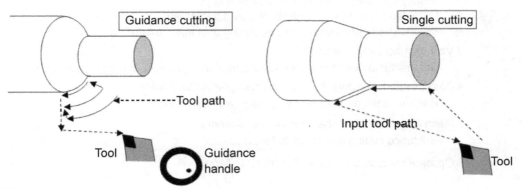

2. Single Cutting

- Tool path such as positioning, line and Circle can be inputted directly.
- Teaching, Editing of taught-in blocks and Playback machining are available.

Module Three CNC Machining

Easy program checking

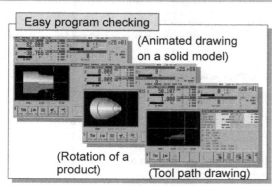

(Animated drawing on a solid model)

(Rotation of a product) (Tool path drawing)

Single cutting

- Tool path such as positioning, line and circle can be inputted directly.
- Teaching and editing of taught-in blocks and Playback machining are available.

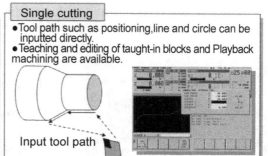

Input tool path

Tool (Single cutting screen)

Cycle cutting

By inputting a machining figure, all of the machining motions can be done automatically.
- Bar machining
  (Outer / Inner / End face)

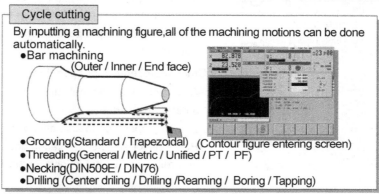

(Contour figure entering screen)

- Grooving(Standard / Trapezoidal)
- Threading(General / Metric / Unified / PT / PF)
- Necking(DIN509E / DIN76)
- Drilling (Center driling / Drilling /Reaming / Boring / Tapping)

New fatures exclusively designed for shop-foor operations(1)

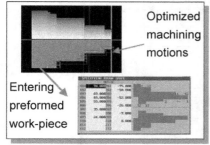

Optimized machining motions

Entering preformed work-piece

(Air Cut Canceling)

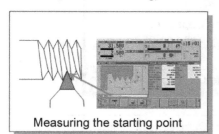

Measuring the starting point

(Thread Repair Cycle)

1. Air Cut canceling function
   - The optimized machining motions can be carried out automatically in accordance to the entered preformed work-piece figure data.
   - Processing time can be reduced.
2. Thread Repair Cycle
   - The starting point of threading can be measured by making a threading tool touch onto the actual thread.
   - A broken thread can be repaired easily.
3. Necking cycles
   - DIN509-F necking pattern can be input easily in addition to DIN509-E and DIN76.
4. Adaptation to work coordinate(G55-G59)
   - G55-G59 work coordinate can be selected in addition to G54.

New Features exclusively designed for shop-floor operations(2)

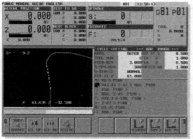
(Enlarging input contour)

(Tool File window)

5. Addition of contour enlarging function
   - Detail of contour can be checked easily by enlarging.

6. Addition of a Process List for selecting a executing process
   - Only finishing as the need arises after measuring can be executed.

7. Implementation of explanatory diagram
   - Diagram explaining a position of imaginary tool nose was added.

8. Other improvement for operation
   - A function to prevent selecting a wrong machining program was added.
   - Zooming operation in animation screen was optimized.
   - Positions for spindle orientation and the second reterrence can be set in initial screen.

FANUC MANUAL GUIDE i for Compound Machine

Operation Guidance for Compound Machine

Compound Machine

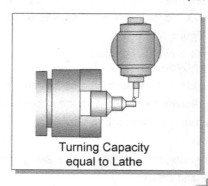
Turning Capacity equal to Lathe

+

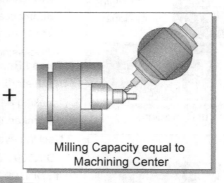
Milling Capacity equal to Machining Center

High efficient machining, but complicated operation

Simple operation with MANUAL GUIDE for Compound Machine

Accomplish simple operation on a Compound Machine

Module Three CNC Machining

(Example of making a program)

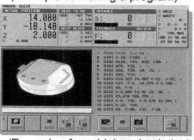

(Example of machining simulation)

Applicable CNC : FANUC Series 16*i*/18*i*-TB
　　　　　for compound machining function

- Relief of the complicated operation
  - Display and operations were integrated to one screen.
    Offset data, Work coordinate system
    Current position, G/M/S/T-code display
- Easy making of machining program
  - Operator can make program easily in dialogue form without knowledge for turning and milling.
- Easy checking of machining program
  - Realistic drawing both of turning and milling with 3-D solid model is available.
  - Milling on a slanted surface can be simulated.
  - Cutter mark according to a tool tip shape can be expressed.

Support whole machining and turning facilities

milling Cycles

(Tool path of Contour Pocketing)

- Contour machining(Side wall, Pocket, Groove)
- Hole machining(Center drilling, Drilling, Reaming, Boring, Tapping)
- Hole pattern(Points, Line, Arc, Circle, Square, Grid)
- Facing(Square, Circle)
- Side cutting(Square, Circle, Track, One side)
- Pocketing(Square, Circle, Track, Groove)
  Note:All of the above can be done also on a slanted surface
- Measuring(Centering/Product measurng for turning.milling)

Turning Cycles

(Tool path of Bar Roughing)

- Bar machining(Roughing)
- Bar machining(Finishing)
- Threading
- Grooving(Normal, Trapezoidal)
- Hole machining for lathe(Center drilling, Drilling, Reaming, Boring, Tapping)
- C-axis hole machining
- C-axis grooving
- C-axis cylindrical machining

Module Four

Die & Mold Tech

Scene One   Transfer Molding

Fill in the blanks.

_____ molding represents a further development of _____ molding. The uncured thermosetting _____ is placed in a heated transfer pot or _____. After the material is heated, it is _____ into heated closed molds.

 Survey your workplace, and try to describe the sequence of operations in transfer molding for thermosetting plastics.

Look through the following schematically illustration, please try to fill in the blanks about which stage of each picture describes after you write down the terms the figures refer to and then put the figures in order according to the sequence of operations in transfer molding for thermosetting plastics.

Module Four Die & Mold Tech

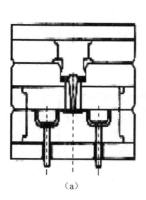
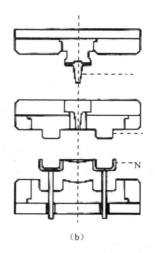
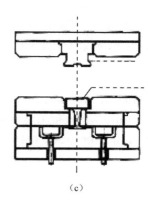

Figure 4-1

(a) _____.

(b) _____.

(c) _____.

Typical parts made by transfer molding are electrical and electronic components and rubber and silicone parts. This process is particularly suitable for intricate parts with varying wall thickness.

Work in your workplace and find more parts that are made by transfer molding with your team mates, then write them down.

_____
_____
_____

Practice spoken English.

Mr. Zhang: Hello, Mr. Cheng, what are you doing?
Mr. Cheng: I am doing some exercises .
Mr. Zhang: What are they?
Mr. Cheng: It is about the kinds of steels.
Mr. Zhang: Do you know the kinds of steels?
Mr. Cheng: Yes, they are ⋯
Mr. Zhang: Is it important in machining?

机械制造类专业英语

Mr. Cheng: Yes.
Mr. Zhang: Why?
Mr. Cheng: …
Mr. Zhang: Thank you for your help.
Mr. Cheng: My pleasure.

New words and expressions we learn in this session(let the students copy the new words and expressions he or she doesn't know in this session and then dictate them among his/her group members)

Grammar 14　数量词的翻译

为使文章紧缩，电子信息类英文中总是尽可能多地使用数量词。英语在数量词的表达上有不同特点，翻译时要持谨慎态度，哪怕不小心错译一位数字，也可能酿成严重后果。译后对数字部分要反复校核，使其准确无误。

数量词翻译要点例举如下。

1. 英语数字表达差异

英语中无"万"单位，要以"千"累计。

　　a ten thousand　　　　　　一万（十千）
　　a hundred thousand　　　　十万（百千）

2. 百万以上数量词

百万以上的大数量词分为大陆制（thousand system，美国、法国和欧洲大部分国家采用）和英国制（million system，英国和德国采用）。汉译这种数字首先搞清不同的制，其主要区别如下。

制 数　字	英　国　制	大　陆　制
109（十兆）	a thousand million	a billion
1012（兆）	a billion	a trillion
1015（千兆）	a thousand billion	a quadrillion

3. 数量单位

　　120V　　　　　　　　　　120伏
　　200℃　　　　　　　　　20摄氏度
　　550 kHz　　　　　　　　550千赫

但有些词并非数量单位，如 bit, byte, digit, phase 等。

　　8 bit　　　　　　　　　　8位字

Module Four　Die & Mold Tech

three-phase poser	三相电

以上用"-"连接的词为形容词，下列为名词。

4 inputs	4 输入
Page 1321	第 1321 页
1993 years	1993 年
1072 Main Street	曼恩大街 1072 号

4．斜线和分数

5/8	八分之五
1/8 inch	八分之一寸
3-11/16	三又十六分之十一
two-thirds voltage	三分之二伏
one-millionth of a farad	百万分之一法（拉）

5. Increase 和增加

翻译中的"n-1"规则适用于以下句子。

Increase by a factor, increase….

The drain voltage has been increased by a factor of five.
漏电压增加了 4 倍。

The production of various electron tubes has been increased four times as against 1958.
各种电子管产量比 1958 年增加了 3 倍。

以下句子不适于"n-1"译法。

A is five times bigger than B.
A 比 B 大 5 倍。

This year the value of our industrial output has increase by twice as compared with that of last year.
今年我国工业产值比去年增加了 2 倍。

A temperature rise of one degree centigrade raises the electric conductivity of a semiconductor by 3%~6%。

另 double, treble, quadruple… 译为"1，2，3，……倍"。

As the high voltage was abruptly trebled, all the valves burnt.
由于高压突然增加了 2 倍，电子管都烧坏了。

6. Reduce 和减少

Cost of radio receivers was reduced by 80%.
收音机成本降低了 80%。

The length of laser tube was reduced ten times.
激光管长度缩短了十分之九（为十分之一）。

This kind of film is twice thinner than ordinary paper.
这种薄膜的厚度只是普通纸张的一半。

127

The members have decreased to 50.
人员减少到 50 名。

Switching time of the new type transistor is shortened three times.
新型晶体管开关时间缩短为三分之一（可缩短三分之二）。

Scene Two　　Piercing and punching die

The punch force builds up rapidly during shearing because the entire thickness is sheared at the same time. Please write down the terms of the following examples of the piercing processes.

　　（a）　　　　　　（b）　　　　　　（c）　　　　　　（d）

Figure 4-2

Activity: Make a survey about piercing die in your workplace.

Task One

Examine your own workplace and make out different piercing die processes.

A complete press tool for cutting a hole or multi-holes in sheet material at one stroke of the press, as classified and standardized by a large manufacture as a single-station piercing die is common in workplace.

Any complete press tool, consisting of a pair (or a combination of pairs) of mating members for producing pressworked (stamped) parts, including all supporting and actuating elements of the tools, is a die. Pressworking terminology commonly defines the female part of any complete press tool as a die.

The guide pins, or posts, are mounted in the lower shoe. The upper shoe contains bushings which slide on the guide pins. The assembly of the lower and upper shoes with guide pins and bushings is a die set. Die sets in many sizes and designs are commercially available.

Module Four　Die & Mold Tech

　　A punch holder mounted to the upper shoe holds two round punches (male members of the die) which are guided by bushings inserted in the stripper. A sleeve, or quill, encloses one punch to prevent its buckling under pressure from the ram of the press. After penetration of the work material, the two punches enter the die bushings for a slight distance.

　　The female member, or die, consists of two die bushings inserted in the die block. Since this press tool punched holes to the diameters required, the diameters of the die bushings are larger than those of the punches by the amount of clearance.

　　Since the work material stock or workpiece can cling to a punch on the upstroke, it may be necessary to strip the material from the punch. Spring-loaded strippers hold the work material against the die block until the punches are withdrawn from the punched holes. A workpiece to be pieced is commonly held and located in a nest composed of flat plates shaped to encircle the outside part contours. Stock is positioned in dies by pins, blocks, or other types of stops for locating before the downstroke of the ram.

 Task Two

　　Teamwork, discuss with your classmates and try to translate the underlined sentences above into Chinese.

 Say out the blanks of the picture.

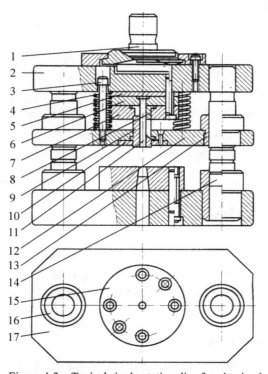

Figure 4-3　Typical single-station dies for piercing hole

机械制造类专业英语

Pair work

One student read aloud the new terms and expressions we learn today, the other one will write them down, and then the first student will translate them into Chinese. Then they exchange their roles and do the activity again.

New words and expressions we learn (let the students copy the new words and expressions he or she doesn't know in this session and then dictate them among his/her group members)

Practice spoken English.

Mr. Li: Excuse me, may I ask you a question? Bob.
Bob: Of course.
Mr. Li: Do you know about the plastic?
Bob: Yes. I know some of them.
Mr. Li: Is plastic important in manufacture industry?
Bob: Plastic is the foundation of machine.
Mr. Li: What's this then?
Bob: It's …
Mr. Li: What's that?
Bob: …
Mr. Li: What's that then?
Bob: Do you know more information?
Mr. Li: Yes. You are very kind.
Bob: Delighted to have been of assistance.

Grammar 15　定语从句

定语从句在句子中作定语，它修饰句子中的某一名词或代词，引导定语从句的关联代词有 who, whose, which, that；关系副词有 when, where, why 等。

关系代词和关系副词除用来引导定语从句外，还可代表从句修饰的先行词，在从句中作主语、宾语、定语和状语。

1. 关系代词和关系副词在从句中的作用

A small business or company may employ only one bookkeeper who records all of the financial data by hand.

Module Four Die & Mold Tech

一个小的企业或公司可以只雇佣一个薄记员用手工操作来记录全部的财务数据。（关系代词做副词）

If the accounts are not balanced, there must be some error which the bookkeeper must find and correct.

如果账户不平，则准有一些错误需薄记员去发现并加以纠正。（关系代词作宾语）

I want to see the boy whose father is a director.

我想见那个父亲是董事长的男孩。（关系代词作宾语）

The girl whom you met at the station is the bookkeeper of this company.

你在车站遇见的那个女孩就是这家公司的薄记员。（关系代词作宾语）

Economics is the science that deals with human wants and their satisfaction.

经济是处理人类供求关系的科学。（关系代词作宾语）

The only reason why we desire goods or service is to satisfy our wants.

我们需求商品或劳务的唯一理由是为了满足我们的需要。（关系副词作状语）

The retail store where I often buy goods is operated by his father

我经常记得在大学学习的那些日子（关系副词作状语）

关系代词及关系副词在从句的作用可归纳如下。

关系代词 { who 指人，在从句中作主语。
whom 指人，在从句中作宾语，在限定性定语从句中一般省略。
which 指物，在从句中作主语和宾语。
that 指物，也可指人，在从句中作主语或宾语。

关系代词 { when 指时间，在从句中作状语。
Why 指原因，在从句中作状语。
Where 指地点，在从句中作状语。

2．限定性定语从句和非限定性定语从句

限定性定语从句是指先行词不可缺少的定语，它与主语的关系十分密切，若将这种定语从句省去，则主句的意思就不完整，且这种定语从句不可用逗号与主句分开。

非限定性定语从句是先行词的附加说明，即使省去了也不影响主句的意思，因此常用逗号把它与主句分开。如：

A financial statement is the statement of changes in financial position, which shows an increase or decrease in working capital for the year.

财务报表是财务状况变化的报表，它显示年度内营运资本的增加和减少。（非限定性定语从句）

The financial statement is the most important statement which provide information about the current financial condition of an enterprise to individuals, agencies, and organzations.

财务报表是向两个人，政府机构和有关组织提供有关一个企业当期财务状况最重要的报表。（限定性定语从句）

3. 关系代词 whom, which 在定语从句中用作介词的宾语。如：

The man to whom you spoke is my teacher.
与你谈话的那个人是我老师。
（通常把 whom 省去，介词 to 仍放在 spoke 后面）
We visited the house in which he once lived.
我们参观了他曾住过的房子。
Double-entry is a method of bookkeeping in which the two-fold effect of every entry is recorded.
复式记账是一种记录每笔分录双重结果的簿记方法。

Scene Three　Injection Molding

Discuss with your team mates about injection molding.　　　　　　True　False

（1）Injection molding is principally used for the production of the thermoplastic parts, although some progress has been made in developing a method for injection molding some thermosetting materials.　　□　□

（2）The problem of injecting a melted plastic into a mold cavity from a reservoir of melted material has been extremely difficult to solve for thermosetting plastics which cure and harden under such conditions within a few minutes.　　□　□

（3）The principle of injection molding is quite similar to that of die-casting.　　□　□

（4）The process consists of feeding a plastic compound in powdered or granular form from a hopper through metering and melting stages and then injecting it into a mold.　　□　□

（5）After a brief cooling period, the mold is opened and the solidified part ejected.　　□　□

（6）Several methods are used to force or inject the melted plastic into the mold. The most commonly used system in the larger machines is the in-line reciprocating screw.　　□　□

（7）The screw acts as a combination injection and plasticizing unit. As the plastic is fed to the rotating screw, it passes through three zones as shown: feed, compression and metering.　　□　□

（8）After the feed zone, the screw-flight depth is gradually reduced, forcing the plastic to compress.　　□　□

（9）The work is converted to heat by shearing the plastic, making it a semifluid mass. In the metering zone, additional heat is applied by conduction from the barrel surface.　　□　□

Injection-molding machines can be arranged for manual operation, automatic single-cycle operation, and full automatic operation. The following are the advantages of injection molding. Can you translate them into Chinese with your partners?

1. A high molding speed adapted for mass production is possible.

Module Four Die & Mold Tech

2. There is a wide choice of thermoplastic materials providing a variety of useful properties.

3. It is possible to mold threads, undercuts, side holes, and large thin sections.

Write down the terms in the following figures and then work in your workplace and try to find out the meaning of the terms in the following figures. Then work with your group members, describe and dictate the new terms for each.

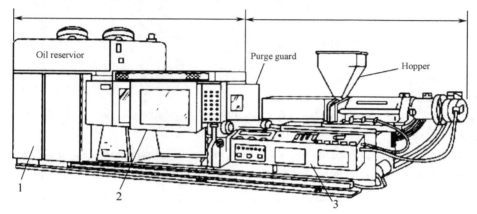

Figure 4-4   The injection-molding machine

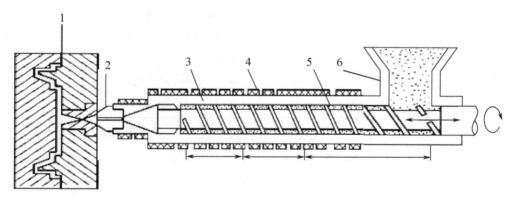

Figure 4-5   The reciprocating-screw injection system

New words and expressions we learn in this session (let the students copy the new words and expressions he or she doesn't know in this session and then dictate them among his/her group members)

Practice spoken English.

Miss Li: Hi, Mr. Liu.

Mr.Liu: Hi, Miss Li, how are you?

Miss Li: I am fine, thank you, but I have some problem with hot working and heat treatment. Could you give me a favour?

Mr.Liu: Yes, my pleasure. What are these problems?

Miss Li: I don't know how they works?

Mr.Liu: So you can look at the reference book in the library .

Miss Li: Do you know heat treatment?

Mr.Liu: …

Miss Li: Is hot working used in mass production?

Mr.Liu: Yes.

Miss Li: Why?

Mr.Liu: Because …

Miss Li: But, for mass production, what other methods should we use?

Mr.Liu: …

Miss Li: I am really appreciated you for telling me these knowledge.

Mr.Liu: I am glad to be of some services.

Grammar 16　派生词的翻译

电子信息专业英语词汇有相当一部分以派生词法构成。所谓派生词法，这里指词根加上前缀和后缀形成的词。大部分词缀都源于拉丁语和希腊语。掌握词缀及其派生词的翻译要点，对迅速准确地译出电子信息类专业原文有很大的帮助。

本单元和下单元的翻译技巧将分别介绍电子信息工作人员必须掌握的主要词缀及其汉译。

1. 形容词词缀

词缀	译义	词例	译义
in-	不，非	insufficient	不足的
im-	不，非	imompatible	不兼容的
un-	不，非	unstable	不稳定的
super-	上，超	supersonic	超音速的
-ive	…的	reactive	电抗性的
-ar	…状的	linear	线性的
-ic	具有…性质的	electronic	电子
-ous	有…特性的	synchronous	同步
-proof	防…的	fireproof	防火

2 动词词缀

词缀	译义	词例	译义
re-	再	reuse	再使用
over-	过分	overload	过载
under-	不足	underload	欠载
dis-	相反	discharge	放电
de-	取消	demagnetize	去磁
trans-	横穿	transform	变换
-fy	使…	electrify	使带电
-ize	…化	normalize	正常化
-ate	使…形成	integrate	使成一体

词缀	译义	词例	译义
inter-	相互在…中间	interface	界面 接口
counter-	反对应	counterpart	对应部分 副本
in-	在…内部	inlet	入口
out-	在外	output	输出
di-	二	diode	二极管
tele-	远	teleconference	远程电话会议
photo-	光	photosphere	光球
micro-	微小	microwave	微波
ultra-	超	ultrahigh frequency	超光频

Scene Four Bending Die

 Can you name some of the terminology used in bending?

 Please translate the words in the following figure into Chinese.

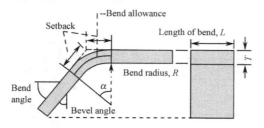

Figure 4-6

机械制造类专业英语

 Write down the new words and expressions in the contents above.

Fill in the blanks.

Minimum bend radius vary for _____ metals; generally, different annealed metals can be _____ to a radius equal to the _____ of the metal without _____ or weakening. As R/T decreases (the ratio of the bend radius to the thickness becomes smaller), the _____ strain at the outer fiber _____, and the material eventually _____.

The minimum bend radius for various materials at room temperature are given in the following table.

Material		Condition	
		Soft	Hard
Aluminum alloys		0	6T
Beryllium copper		0	4T
Brass, low-leaded		0	2T
Magnesium		5T	13T
Steels	Austenitic stainless	0.5T	6T
	Low-carbon, low-alloy, and HSLA	0.5T	6T
	Titanium	0.7T	4T
	Titanium alloys	2.6T	4T

Note: T——thickness of material

An approximate formula for the bend allowance, L_b is given by

$$L_b = \alpha (R + kT)$$

Fill in the blanks with the corresponding signs in the formula above.

(　) — is 0.33 when R is less than 2T and is 0.50 when R is more than 2T
(　) — the inside radius of bend, in (mm)
(　) — the sheet thickness, in (mm)
(　) — the bend angle (in radians)
(　) — bend allowance, in (mm)

Look at the figures below, point out the bending methods they used.

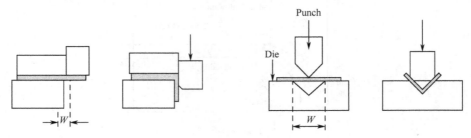

Figure 4-7

Module Four Die & Mold Tech

 Survey your workplace and try to find out more bending methods. Look at the following figures; please do your best to understand the meaning of the terms, and dictate for each other, then give each of the following figure a topic.

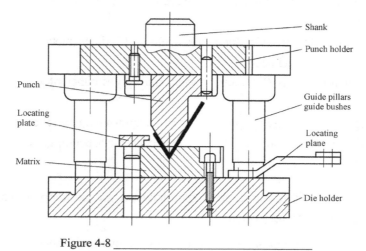

Figure 4-8 _____

Please write down the terms of the following figures with your colleagues.

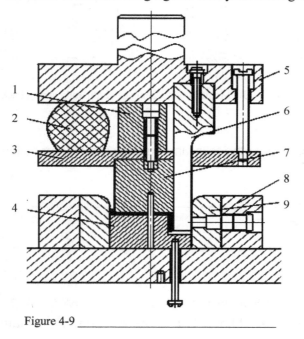

Figure 4-9 _____

 The calculation of bending force is as follows.

$$P=KLST^2/W$$

Fill in the blanks with the corresponding symbols in the formula above.

(　　) — metal thickness, in.

(　　) — width of V or U die, in.

(　　) — ultimate tensile strength, tons per square in.

(　　) — length of part, in.

(　　) — die opening factor: 1.20 for a die opening 16 times metal thickness, 1.33 for an opening of eight times metal thickness

(　　) — bending force, tons (for metric usage, multiply number of tons by 8.896 to obtain kilonewtons)

For U bending (channel bending) pressures will be approximately twice those required for V bending, edge bending requires about one-half those needed for V bending.

All materials have a finite modulus of elasticity, when bending pressure on metal is removed, some elastic recovery occur.

Methods of reducing or eliminating spring back in bending operations are over benting through an angle equal to the spring back angle with an undercut or relieved punch.

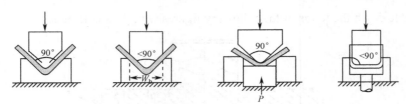

Figure 4-10　Methods of reducing or eliminating spring back

Group activities. Answer the following questions with your colleagues.

What's spring back?

___

How to solve this problem?

___

New words and expressions we learn（let the students copy the new words and expressions he or she doesn't know in this session and then dictate them among his/her group members）

 Practice spoken English.

Mr.Zhang: Good morning, Mr Cheng.
Mr.Cheng: Good morning, Mr Zhang.
Mr.Zhang: What are you doing?
Mr.Cheng: I am machining a work-piece.
Mr.Zhang: What's this?
Mr.Cheng: It is a fixture.
Mr.Zhang: What can it do?
Mr.Cheng: We can use it to locate and hold the work-piece when it is machined.
Mr.Zhang: Is the fixture only be used to machine?
Mr.Cheng: No. In fact there are many kinds of fixture and they can also be used for inspection, welding and assembly.
Mr.Zhang: What are its advantages?
Mr.Cheng: We can improve the product precision and reduce time assumption of locating the work-piece.
Mr.Zhang: It's very kind of you for telling me about these.
Mr.Cheng: That's all right.

Grammar 17 状语从句

在句子中作状语的从句叫状语从句，它可以修饰主句中的动词、形容词和副词。

状语从句根据其含义可分为时间、地点、原因、目的、结果、方式、让步、条件等八种状语从句。

1. 时间状语从句

时间状语从句的引导词有 when, as, while, after, before, since 等。如：

When an expense is incurred, either the assets are reduced or the liabilities are increased.
当一项费用发生后，既是资产减少又是负债增加。

You must listen attentively and take notes while I explain the text.
当我解释课文时你必须注意听并做好笔记。

Before a transaction can be entered properly, it must be analyzed in order to determine which accounts are affected by the transaction.
在完整地记录一笔经济业务之前，必须进行分析，以确定哪些账户受经济业务的影响。

2. 地点状语从句

地点状语从句的引导词有 where, wherever。如：

Make a mark where you find an error.
在你发现有错的地方做个记号。

Wherever we go, we must build up good relations with the customers.
无论我们到哪里，都要与顾客搞好关系。

3．原因状语从句

引导原因状语从句的词有 because, since, as not that 等。如：

The revenue and expense accounts are called temporary owner's equity accounts because it is customary to close them at the end of each accounting period.
收入和费用账户被称为临时性的业主权益账户，因为通常在每个会计年度末要把他们结清。

Since an investment is not classified as income or expense ta the business, it is not considered in the income statement.
由于一项投资不是作为企业的收入或费用来分类的，因此在收益表里不考虑它。

4．目的状语从句

引导目的状语从句的词有 that，so that, in order that 等。如：

We must work harder harder so that we may fulfil our plan ahead of schedule.
我们必须努力工作，以便能提前完成计划。

A separate account is kept for each asset, liability, and capital item so that information can be record for each of them.
要为每个资产、负债、资本项目设备自的账户，以便记录它们的信息。

I get up early every day in order that I may have time to study English.
我每天起得很早，以便有时间学英语。

5．结果状语从句

引导结果状语从句的词有 so… that, so that 等。如：

The coat was so expensive that I could not afford it .
这件外套非常贵，我买不起。

He received money form customers on account yesterday, so that he can pay for equipment purchased on credit.
昨天他从顾客那里收到欠款，现在他可以支付赊购设备的款项了。

6．表示比较关系的状语从句

引导这类状语从句的词有 that, as, as if 等。如：

He ran as fast as he could.
他尽力快跑。

He never reads as mush as is required of the call.
他从不读课外书。

This novel is more interesting that one.
这本小说比那本更有趣味。

7．让步状语从句

引导让步状语从句的词有 though, although, no matter, whatever, even if, however 等。

Module Four　Die & Mold Tech

Although a wide variety of journals are used in business, the simplest form of journal is a two-column journal.

尽管企业使用的日记账种类繁多，最简单的还是两栏式日记账。

Whatever we do, we must fist think of the collective.

无论我们做什么，我们必须首先想到集体。

We'll start on our journey even if the weather is had .

即使天气不好，我们也要去旅行。

8．条件状语从句

引导条件状语从句的词有 if，unless, suppose, as long as 等，如：

If there is only one item entered in a column, no footing is necessary.

如果栏目中只记了一个项目，就没有必要合计了。

He is sure to come unless he has some urgent business.

他一定会来，除非有急事。

Scene Five　Extrusion

 Discuss with your teammates about extrusion.　　　　　　　　True　False

（1）Cold extrusion, or heated to a temperature which gives optimum results for the conditions applying and the product requirements, i.e. hot or warm extrusion.　☐　☐

（2）In most cases, the metal is in the form of a solid or hollow cylinder and is moved in a containing tool in the direction in which the product emerges from the die aperture.　☐　☐

（3）Most metals can be extruded in one way or another, but obviously some more easily than others.　☐　☐

（4）The common metals that can be extruded at ambient temperatures are lead, aluminium, copper, low carbon steels and a few alloys of these metals. The softer semi-precious metals can also be extruded cold if required.　☐　☐

（5）Traditionally, in hot extrusion a billet of metal is forced through a shaped hole at one end of a container by the action of a ram pressing from the other end.　☐　☐

（6）The container bore is of uniform cross-section and the ratio of its area to that of the product is known as the extrusion ratio.　☐　☐

（7）In this case the extrudate leaves the die through the hollow ram. Although the container bore is normally round it may also be rectangular to permit extrusion of very wide products.

　☐　☐

（8）The main tooling components for the hot extrusion process are the die assembly, the dummy block and the mandrel.　☐　☐

（9）In forward rod and tube extrusion the punch does not suffer much from wear but the compressive stresses are similar to can extrusion punches.　☐　☐

There are four basic extrusion operations, discuss with your group members and write them down in the following the schematic illustration.

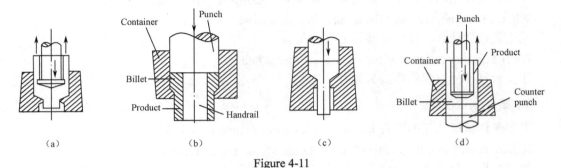

Figure 4-11

Please work with your team mates trying to translate the following sentences into Chinese.

1. Normally dies have to have at least one support ring as the internal pressures are high enough to cause transverse cracking due to tangential or triaxial stresses respectively.

2. In can extrusion the punch is highly stressed by compressive and bending loading, and the same time subjects to heavy wear and increases in temperature at the punch nose.

3. Extrusion is the process in which material is forced by compression to flow through a suitably shaped aperture in a die, usually to give a product of a smaller but uniform cross-sectional area.

Survey your workplace and do your best to observe the different kinds of punch in extrusion, and then give out the different terms of the schematic illustration.

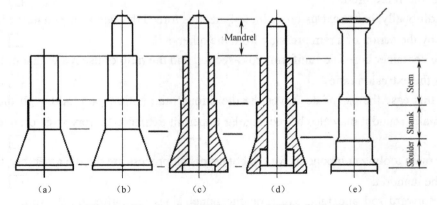

Figure 4-12   Punch and punch /mandrel configurations

Module Four　Die & Mold Tech

 New words and expressions we learn （let the students copy the new words and expressions he or she doesn't know in this session and then dictate them among his/her group members）

 Practice spoken English.

John: How do you do! My name is John.
Miss Li: How do you do!
John: Are there any vacancies in your factory?
Miss Li: Yes. We need some new workers. What was your major?
John: I majored in mechanical manufacturing, and I think I am good at operating milling machine.
Miss Li: What sort of job do you like to do?
John: I would like to be amilling worker. I think I can do my best to be a milling worker.
Miss Li: Great! What time can you start to work?
John: Anytime.
Miss Li: Very nice. How much do you expect in salary?
John: 800 Yuan.
Miss Li: OK! Please fill in this form. If anything turns, I'll contact with you.
John: Thank you!
Miss Li: You're welcome.

Grammar 18　翻译的两个过程

翻译过程可分为两个阶段，即原文理解过程和译文表示过程。

1. 原文理解过程

翻译的前提是对原文的正确理解。有了这个前提，才能运用译文的语言规律来表达原文的意思。原文理解过程可归纳为3步：通读全文，领略大意；仔细推敲，辨识词义；分析语法，明晰关系。

1）通读全文，领略大意

我们知道，英语行文时整个语篇是有其总体构思的。而句子所表达的意思与其总体构思密不可分。因此，在逐字逐句地翻译原文之前，通读全文，领略整篇大意是很有必要的。

2）仔细推敲，辨识词义

表达思想采用的基本语言单位是词，句子就是按一定的语言习惯、语法规则和修辞手法，由许多词组合起来组成的。因此，不能确切地了解句中每个词的含义，就不能正确理解全句的意思。

那么，怎么辨明词义呢？下面是应注意的几个方面。

A. 查字典

翻译中，遇到词义难以确定的词，切不可望文生义，一定要多查字典。特别要注意一词多义，应仔细推敲，弄清词义。如有这样一个标志牌，上面写着：

marketing package

marketing 这个词，通常人们理解为"市场"。但此处若这样理解就错了，它还有一个意思是"销售"，因此译文应是：销售包装

B．根据上下文，在整句话中理解词义

英语词义比较灵活，词的含义范围较宽，词义多变，词义对上下文的依赖性比较强，只有结合上下文才能确定下来。或者说英语的词义，只有在整句话的整体意思上才有具体的含义。

C．从专业知识角度分析

在科技英语翻译中，对一些词义不明确的词，除了结合上下文，还要联系专业内容来确定其具体含义。例如：

薄膜类包装　Film-Based packaging

如果不结合专业知识，误将已专业化了的词汇 film 仍然按普通词汇的含义来翻译，则可能会被误译为：基于电影的包装

但若结合专业知识就知道，film 的意思是"薄膜"，这样便得到正确译文了。

3）分析语法，明晰关系

要理解原文，除了弄清词义，还要根据原文的句子结构，搞清这些词之间的语法关系，也就是说，要按照原语的语法规则，认识句中各成分之间的关系。具体说来，首先弄清整句的脉络，搞清是简单句还是复合句。若是简单句，则确定主语、谓语、其他附属成分和修饰语；若是复合句，则确定其为何种复合句，然后分别按简单句的方式，对句子进行分析。

2．译文表达过程

译文的表达过程，归纳为三稿定案：一稿初译，忠实为主；二稿推敲，力求准确；三稿定案，讲究修辞。

1）一稿初译，忠实为主

首先，根据原文的特点决定采用直译法还是意译法。若一时把握不准，可先采用直译，待以后再做进一步考虑，这样做易于忠实原文。

2）二稿推敲，力求准确

一稿出来后，不管是用哪种方法翻译，其译文毕竟还是初稿，因此，应多方推敲，力求准确。推敲所选的词，其词义是否确切；词的搭配是否得当；所造的句子，是否会引起歧义，是否符合汉语的语法规范和语言习惯等。

3）三稿定案，讲究修辞

虽然科技英语本身修辞方式不多，但汉语毕竟是一种讲究修辞的语言。因此，在最后敲定译文前，还应仔细琢磨，看是否要采用某些汉语的修辞手法，如词的增减和重复、适当使用连词、灵活安排句式等。

Scene Six　Drawing Die

Drawing is a process of changing a flat, precut metal blank into a hollow vessel without excessive wrinkling, thinking, or fracturing. The various forms produced may be cylindrical or

Module Four Die & Mold Tech

box-shaped with straight or tapered sides. Many parts made of sheet metal are cylindrical or box-shaped: for example, pots and pans, containers for food and beverages, kitchen sinks, and automotive fuel tanks. Such parts are usually made by a punch forces a flat sheet-metal blank into a die cavity. Although the process is generally called deep drawing, on account of its capability of producing deep parts, it is also used to make parts that are shallow or have moderate depth. The following figure can help you understand drawing process.

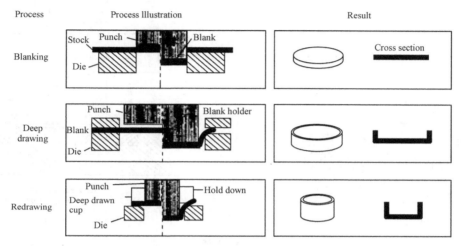

Figure 4-13 The metal-forming process

Can you name the corresponding Chineseterms?

 Please tick out the items if it can be made by drawing die .

paper clip（ ）　　　file cabinet（ ）　　flanges（ ）　　seams（ ）
corrugations（ ）　　cylindrical（ ）　　box（ ）　　　cup（ ）
pots（ ）　　　　　　pans（ ）　　　　　containers（ ）
kitchen sinks（ ）　　automotive fuel tanks（ ）

Write down the terms about the following figure.

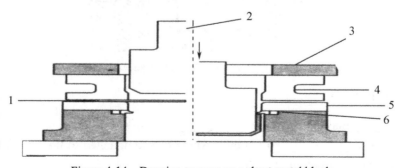

Figure 4-14 Drawing process on a sheet-metal blank

👉 The following is a simple type of draw die in which the precut blank is placed in the recess on top of the die, and the punch descends, pushing the cup through the die. Can you name the terms in Chinese? The second figure shows a simple form of drawing die with a rigid flat planholder for use with 13-gage and heavier stock. Discuss with your classmates the terminology and try to remember them in mind.

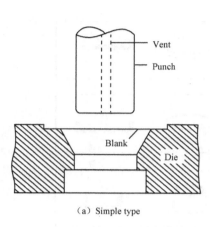 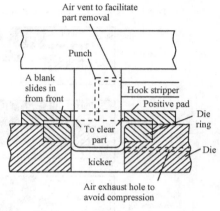

(a) Simple type   (b) Simple draw die for heavier stock

Figure 4-15

👉 Look over the following figure and discuss with your partners the process it conveys, and then translate the terminology for each other. Please write down the terms of the following figure.

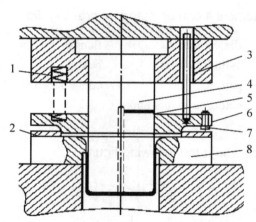

Figure 4-16 First drawing die with elastic blank holder

✎ Listen and fill in the blanks.

Deep _____ may be made on _____ dies, where the _____ on the blank holder is more evenly _____ by a die cushion or pad attached to the bed of the press. The following is a typical construction of such a _____ .

Module Four　Die & Mold Tech

 New words and expressions we learn（let the students copy the new words and expressions he or she doesn't know in this session and then dictate them among his/her group members）

 Practice spoken English.

> Mr.Zhang: Hello, Mr. Chen.
> Mr.Chen: Hi, Mr. Zhang.
> Mr.Zhang: I have some trouble with the primary motion. Could you like to help me?
> Mr.Chen: Yes, I would like to do it. What are these problems?
> Mr.Zhang: How can we distinguish the kinds of motion?
> Mr.Cheng: We can distinguish them by …
> Mr.Zhang: Oh.
> Mr.Chen: So primary motion is …
> Mr.Zhang: That's right .
> Mr.Chen: And feed motion is …
> Mr.Zhang: You are right.
> Mr.Chen: So the differences are …
> Mr.Zhang: Oh, I see. Thank you for telling me these.
> Mr.Chen: You are welcome.

Grammar 19　and 引导的句型

1. and 表示对照

例：Motion is absolute, and stagnation is relative.
运动是绝对的，而静止是相对的。

2. and 表示结果

例：Operators found that the water level was too low so they turned on two additional main coolant pumps, and too much cold water flowing into the system caused the steam to condense, further destabilizing the reactor.
操作人员发现冷却水的水位过低，就启动了另外两位主冷却汞，结果过量的冷却水进入系使蒸汽冷凝，反应堆因而更不稳定。

3. and 表示条件

例：Even if a programmer had endless patience, knowledge and foresight, storing every relevant detail in a computer, the machine's information would be useless, and the programmer know little how to instruct it in what human beings refer to as commonsense reasoning.
即使一个一个拥有耐心知识和眼光的程序员，让所有相关数据都输入计算机，那么机器的信息仍是无用的。因为一个程序员往往因为人类的一些常识原因而不知如何去命令它。

4．and 表示递进

例：The electronic brain calculates a thousand time quicker, and more accurately than is possible for the human being.

计算机在速度和精确度上要比人类快捷千倍。

5．and 表示转折

例：While industrial laser today are most often used for cutting, welding, drilling and measuring, and the laser's light can be put to a much different use: separating isotopes to produce nuclear fuel.

虽然今天的工业激光器经常用于切割、焊接、钻孔和测量，但激光也可以有其他不同的用处，即分离同位素以生产核燃料。

6．and = with

例：In addition, a plug-in fault diagnostic unit and signal monitoring system is normally supplied with the drive to enable any drive to alarm or control signal to be checked and monitored.

此外，一个带有信号监控系统的插入式故障诊断器通常装有驱动装置，以便能校对可监控任何驱动报警信号或控制信号。

Scene Seven Compression Molding

 Read and do your best to understand the following short passage with your partners.

In compression molding, a reshaped charge of material, a premeasured volume of powder, or a viscous mixture of liquid resin and filler material is placed directly into a heated mold cavity. Forming is done under pressure form a plug or from the upper half of the die (Figure 3-1). Compression molding results in the formation of flash (if additional plastic is forced between the mold halves ,because of a poor mold fit or wear, it is called flash.),which is subsequently removed by trimming or other means.

Which of the following are made by compression molding? Please tick it out.

☐ dishes　　　　☐ handles　　　　☐ container caps　　　　☐ fittings
☐ electrical and electronic components　　　　☐ washing-machine agitators
☐ housing　　☐ fiber-reinforced parts with long chopped fibers

There are three types of compression molds in workplace. Please match the relevant items.

semi positive　　　　　　　　　　　　for shallow or flat parts
positive　　　　　　　　　　　　　　for quality production

Module Four Die & Mold Tech

flash-type   for high density parts

Watching the following figures carefully, then try to give each of the following figures a topic.

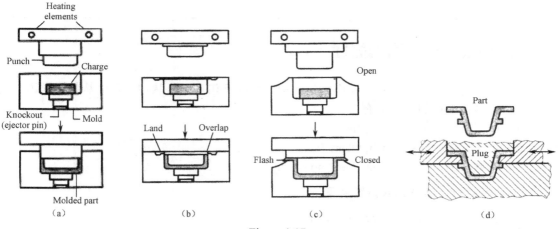

Figure 4-17

(a) _____

(b) _____

(c) _____

(d) _____

 New words and expressions we learn in this session (let the students copy the new words and expressions he or she doesn't know in this session and then dictate them among his/her group members)

Practice spoken English.

> Mr. Cheng: Good morning, Mr. Zhang.
> Mr. Zhang: Good morning, Mr. Cheng.
> Mr. Cheng: Could you help me?
> Mr. Zhang: What can I do for you?
> Mr. Cheng: Can you tell me the definition of metal cutting?
> Mr. Zhang: Let me try. (After a while) That's not too difficult to me.
> Mr. Cheng: What's wrong with it?

149

Mr. Zhang: Oh, I look up the book for it.
Mr. Cheng: Oh, I see.
Mr. Zhang: Metal cutting is…
Mr. Cheng: I know it now. Many thanks.
Mr. Zhang: It's my pleasure.

Grammar 20　名词从句

名词从句是指在句子中起名词作用的从句，名词从句有主语从句、表语从句、宾语从句三种。引导名词从句的关联词有：

（1）连词，如：that, if, whether。
（2）关系代词，如：which, what, who。
（3）关系副词，如：when, where, how, why。

1．主语从句，即名词从句做主语

Whether he comes or not does not matter.
他来与不来都无关紧要。
What they want are financial aids.
她们想要经济援助。

有时会把连接词或关系副词引出的主语从句放在句子后面，用 it 作形式上的主语。如：

It does not matter whether he comes or not.
It was obvious that the driver could not control his car.
显然，司机已不能控制他的车了。
(That the driver could not control his car was obvious.)

2．表语从句，即名词从句作表语，位于主句的联系动词之后。如：

The legal reason for maintaining payroll accounting records is that employers are required by federal, state and local laws to do so.
保留工薪会计记录是由于联邦、州、地方政府的法律的要求。
It seems that he has never been paid the money.
看起来他从没得到钱。

3．宾语从句，即名词从句作宾语，它可以是谓语动词的宾语，也可以是动词不定式，分词或动名词的宾语。如：

He is sure that his experiment will succeed.
他坚信他的试验会成功的。
A person perceives what he looks for and very little else.
一个人要了解他追求什么，别的就不必探究。
He often thought of how he could help others.
他经常在想如何去帮助别人。

Module Four　Die & Mold Tech

引导宾语从句的连词 that 有时可以省略。如：
Saying he was busy, he want away in a hurry.
他说他很忙，即匆忙走了。

 ## Scene Eight　Compound Die

 Compound Die Design

Several operations on the same strip may be performed in one stroke at station with a compound die. Such combined operations are usually limited to relatively simple shapes, because they are somewhat slow and because the dies (with increasing complexity) rapidly become much more expensive than those for individual shearing operations.

A typical example is made in following schematic illustrations: (a) before and (b) after blanking a common washer in a compound die. Note the separate movements of the die (for blanking) and the punch (for punching the hole in the washer). (c) Schematic illustration of making a washer in a progressive die.

Give each of the following figures a topic.

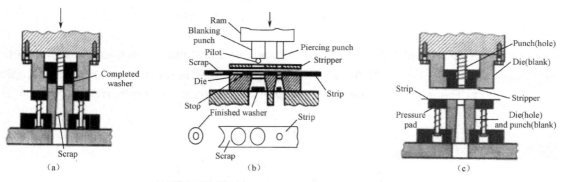

Figure 4-18 _____

Work with your group member discussing the figures above and one spells out the terms, the other translate them into Chinese and then exchange roles to do this exercise.

Team work. Can you describe what compound die is?

Group activity. Work with your group members looking carefully the following figures and discuss how a compound die performs and try to translate the following terminology into Chinese, and then do your best to remember them in mind, lastly dictate for each.

Put the topics in the corresponding place.

A compound blanking and piercing dies, lower shoe assembly, upper shoe assembly.

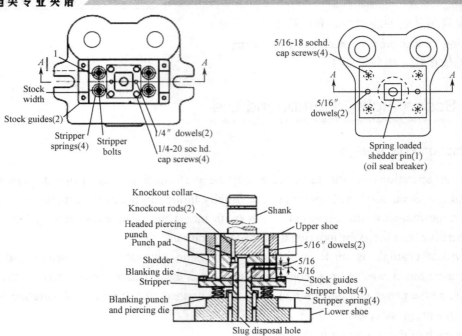

Figure 4-19　A compound blanking and piercing dies

Work with your team mates in your workplace and find more compound dies, do your best to be familiar with more terms as much as possible. The following is a good example for you to study. Please write down the terms with your workmates and then give a topic to it.

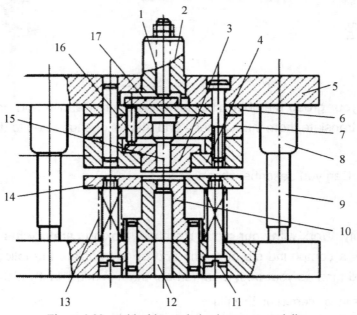

Figure 4-20　A blanking and piercing compound dies

Module Four Die & Mold Tech

 New words and expressions we learn（let the students copy the new words and expressions he or she doesn't know in this session and then dictate them among his/her group members）

Practice spoken English.

Mr.Zhang: Hello, Mr. Chen.
Mr.Chen: Hi, Mr. Zhang.
Mr.Zhang: I have some trouble with the cutting tools. Could you like to help me?
Mr.Chen: Yes, I would like to do it. What are these problems?
Mr.Zhang: How can we distinguish the kinds of cutting tools?
Mr.Chen: We can distinguish them by their shape.
Mr.Zhang: Oh.
Mr.Chen: This is a cutting tool for turning and it have only one point and two edges.
Mr.Zhang: What's this?
Mr.Chen: It is a twist drill. It has two strips of spiral grooves and it usually is used for drilling a hole.
Mr.Zhang: What's that?
Mr.Chen: That is end milling cutter. It is used for machining a plane.
Mr.Zhang: Oh, I see. Thank you for telling me these.
Mr.Chen: You are welcome.

Grammar 21 专业名词的翻译

1. 公司名

公司名一般由两个部分组成：公司名称和表示公司的词，如 company(Co.)，Limited(Ltd.)，corporation(Corp.)，incorporation(Inc.)，另外还有 business，enterprise，organization，establishment，system 等，有些公司名称不带任何"公司"字样，或省略或免用，但仍可根据其性质译为"XX 公司"。

Western Union Telegraph Company	西联电报公司
American Satellite Corporation	美国卫星公司
Northern Radio Company Inc.	北方无线电公司
Central Signals Establishment	中央信号公司
Bharat Electronics Limited	巴拉特电子有限公司

153

公司名前半部是主体，可采用意译。
为从名字上看出公司性质，尽可能意译。

Electronic Tele-Communications, Inc.	电子电信公司
Electronic Tele-Communications, Inc.	电子数据系统公司
Preformed Line Products Company	预制线路产品公司

以人名、地名和难以意译的词命名的可音译。

Bell South Corporation	贝尔南方公司
Fujitsu Limited	富士通有限公司
Summa Four Inc.	萨默 IV 公司

部分公司可直接用缩略名。

AT&T	AT&T 公司
3M	3M 公司
FAC	FAC 公司

2．商品名

商品名的编制虽不免带有一定的随意性，汉译却仍然要持严谨态度，切不可见音音译，见意意译，主观臆测，随手拈来。除必须遵循翻译的一般规律外，还要考虑到商标本身固有的特点，这里提出若干要点。

（1）约定俗成。凡已有的并已认可的商标，汉译一般不要另找译词，沿用即可。

Toshiba	东芝
Siemens	西门子
Motorola	摩托罗拉

（2）注意商标特征。商标设计一般要考虑到 4 个要素：简短易记；发音响亮；具有专利性；译成外文不产生歧义。这些可作为商标汉译的依据。如一个好的商标一般不超过 3 个音节，且首字母发硬音或者含有硬音（C，K，G，Q，X），汉译要考虑原设计的用意。

Casio	卡西欧
Kodak	柯达
Cannon	佳能

（3）用地名、人名、单位名或者其他专有名词做商标的，一般采用原有译词。

Bell	贝尔（大名）
Oxford	牛津（地名）
Cornell	康纳（美国大学名）

（4）体现原名内涵。计算机商标常常带有高技术气息，而通信商标则往往联想到"通"、"远"、"达"等字眼，汉译时要予以考虑。

OPCOM	光通（光通信公司商标）
FITEL	飞达（光纤通信公司商标）

但 Apple 只能译成"苹果"。这是原设计者的意图，目的是要用老幼皆知的水果冲淡苹果高科技的神秘。

Module Four　Die & Mold Tech

Scene Nine　Thermoplastic &Thermosetting Mold Design

Basically, there are two types of transfer molds: the conventional sprue type and the positive plunger type. In the sprue type the plastic performs are placed in a separate loading chamber above the mold cavity. One or more sprues (the runway between the injection machine's nozzle and the runners or the gate) lead down to the parting surface of the mold where they connect with gates to the mold cavity or cavities.

Watch the following schematic illustration of transfer molding carefully and discuss with your group members.

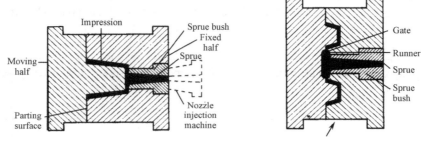

Figure 4-21

Survey your workplace with your classmates carefully and find more in mold. Look at the following figure carefully and then try to write down the terminology used in the figures with your teammates.

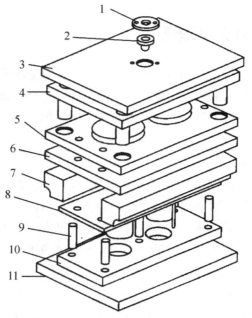

Figure 4-22　The typical structure of two-plate mold

155

机械制造类专业英语

✏️ Translate the following sentences into Chinese with the help of your partners. Then exchange your roles.

1. The molds used for compression molding are classified into four basic types, namely, positive molds, landed positive molds, flash-type molds, and semi positive molds.
___

2. A landed positive mold is similar to a positive mold except that lands are added to stop the travel of a predetermined point.
___

3. In a flash-type mold, flash ridges are added to the top and bottom molds.
___

4. Compression molding is mostly for thermosetting plastics which have to be cured by heat in the mold.
___

5. To speed the process as much as possible, molding presses are usually semi or fully automatic.
___

📖 New words and expressions we learn in this session（let the students copy the new words and expressions he or she doesn't know in this session and then dictate them among his/her group members）
___

Grammar 22　词义辨识法

词是语言中最基本、最活跃的元素，如果译者连词义都无法弄清，就根本没法理解句子的意思。因此，词义辨识法是翻译过程中的第一道关卡，是英语翻译的基本功。

从语义学理论上说，任何一个词都可能具有3个词义范畴，它们是：结构词义，涉指词义和情景词义。

1. 根据词型结构辨识词义

前面我们分析了科技英语在词汇和词法方面的特点，指出了英语词汇，常常根据合成法、词缀法、混成法、缩略法和裁减法构成。因此，熟悉以上构词方法和前述科技英语在词汇和词法方面的特点，学会从词形结构上辨识词义，对快速掌握词汇大有裨益。

比如，根据合成法，可以看出 warehouse 是由 ware 和 house 两个词合成的，因此它的词义自然是仓储。

2. 根据涉指关系辨识词义

根据词形结构辨识词义，是辨识词义的一种有效方法，但这方法却有很大的局限性。这

是因为，有些词在任何情况下，都无法对其进行词形结构的分析。有些词本身并不具备确定的或充分的词义，或者说，这些词的确定的或充分的词义不在其本身，而在其与上下文中的某些词的照应关系中。所谓涉指关系，就是指词在上下文中的对应关系。

3．根据情景关系辨识词义

根据涉指关系辨识词义的局限性也很大，因为很多需要辨识词义的词，在句子或段落乃至整篇文章中都根本不存在什么对应关系。存在对应关系的词项，毕竟是有限的。

有些词需要联系上下文来判断、辨识词义，这就是所谓的根据情景关系辨识词义的方法。

情景，也称为"语境"，它有 3 个范畴：一是指词的联立关系，范围比较小；二是指词的上下或前后文，范围比较大；三是指全段，乃至全篇章所涉及的情节、主题和题材等，范围最大。

1）根据词的联立关系确定词义

英语的词义对上下文的依赖性比较大，独立性比较大。一个词的词义往往取决于与不同词构成的不同搭配或组合。

如 ware 一词，与不同的词建立不同的联立关系时，有不同含义。

warehouse	仓储
houseware	日用品
sofeware	软件
hardware	硬件

2）根据词的上下文或前后文确定词义

在很多情况下，一个词根据相邻关系还没法确定词义，这时需纵观上下文，瞻前顾后，才能依情景分析出词义。

所谓纵观上下文，就是要在上下文中找到判断该次词义的提示，这些提示主要有以下几种。

A．语法功能的提示

英语中，起不同语法作用的词的词性是不同的，而词性不同，其词义便不同。因此可根据词在句中的语法功能的提示来确定词义。如：

The elements like pacing, shiping and distribution are very important.

句中，like 的词义有许多，但从其语法功能上看，此处应为介词，因此，意为"像"。全句翻译为：

像包装，运输和分发是非常重要的。

B．使用场合的提示

英语的词根据出现的场合不同，词义也不尽相同。如 order 一词：

order selecting	分拣
technical order	技术规程
order of connection	接通次序
order for	订单

C．名词数的提示

英语中，有些名词的单数形式和复数形式词义是不同的。

public warehousing	公共仓储	public warehouses	公共仓库

D. 专业内容的提示

同一词汇在不同专业领域内有不同的含义。

Receiving is very important in logistics.

收货在物流中是非常重要的。

Exercises: Answer the following questions.

What does flash mean?
_____

What are the advantages of injection molding?
_____

What is the definition of an undercut?
_____

 Describe the differences between thermosetting and thermoplastic materials.
_____

Describe the structure of two-plate mold.
_____

Translate the following sentences into Chinese.

1. Compression molding results in the formation of flash (if additional plastic is forced between the mold halves, because of a poor mold fit or wear, it is called flash),which is subsequently removed by trimming or other means.
_____

2. Injection molding (injection of plastic into a cavity of desired shape. The plastic is then cooled and ejected in to its final form. Most consumer products such as telephones. Computer casings and CD players are injection molded) is principally used for the production of thermoplastic parts, although some process had been made in developing a method for injection molding some thermosetting materials.
_____

Module Four   Die & Mold Tech

 Translate the following technical terms into English.

注塑模		吹塑模		传递模	
压缩成型		供料仓		加工成本	
内浇口		中心浇口		模具型腔	
成形压力		排气槽		浇注系统	

Translate the following technical terms into English.

板料		冲孔模		冲裁模	
弯曲模		成形模		拉深模	
复合模		挤压模具		热塑性模具	
热固性模具		定位销		定位板	

Self assessments

Time _____ Class_____

No.	Name	Evaluation Items				Notes
		Attitude	Presentation & Passage	Exercises	Activities	

Teachers' Observation Record and Evaluation Sheet

level \ item	listening	speaking	Reading	writing	familiarity of parts
Good					
Better					
Best					
Bad					

New Words

A

a great deal of 大量的
a small portion of 小部分
a small proportion of 小部分
a variety of 种种（若干，各种）
a wide range of 一系列的，一定范围的，一定程度的
ABA Acrylonitrile-butadiene-acrylate 丙烯腈/丁二烯/丙烯酸酯共聚物
abatement 消除
able 有能力的，能干的
able to+动词原形 能，会
abnormal handling 异常处理
abrasion 擦伤，磨损
Abrasion 磨损
abrasive 摩擦（损）的
ABS Acrylonitrile-butadiene-styrene 丙烯腈/丁二烯/苯乙烯共聚物
absolute mode 绝对方式
absolute positioning 绝对（坐标）定位
absorb 吸收，吸引
AC（Alternative Current）交流电流
AC servo motor 交流伺服电动机
accelerator 加速者，加速器，催速剂
acceptable 可接受的
acceptance = receive//n. 验收
accessory 附件，附属设备 adj. 附加的，次要的
accommodate. 供应，供给，使适应，提供，容纳，接纳
accomplish 达到，做到，完成
according to 根据，视……而定
accuracy 精确性，正确度
accurate 正确的，精确的
accurately 正确地，精确地

acid cleaner 酸洗液
acronym
acrylate 丙烯酸盐
ACS（Adaptive Control System） 自动补偿系统
activate 使活动，开动，启动，对……起作用
activated 有活性的
active plate 活动板
active program number 当前运行的程序段号
actuate 开动，驱动，激励
actuator space 作用（工作）空间
actuator 驱动器，执行机构，激励者，激励器
acute 敏锐的，急性的，剧烈的
add lubricating oil 加润滑油
additional. 另外的，附加的，
additive processes 填加材料成形
additive 附加的，添加的，添加剂，添加物
address 地址，地址代码，写地址，从事，忙于
adjacent 邻近的，接近的
adjacent pockets empty 最大刀具直径（相邻空位）
adjust 调整，调节，使适合
administration/general affairs dept 总务部
adoption 采用
advance 前进，推进
advent 出现，到来
advisability 明智
advisable 可取的，明智的
aeronautics 航空学

New Words

AES　Acrylonitrile-ethylene-styrene 丙烯腈/乙烯/苯乙烯共聚物
affordable 可承担起的，可支付得起的，有能力做的
aggregate 组合的
aggregate machine tool 组合机床
aggregation 聚集，集合体
aggressive 好斗的，侵略性的
agitator 搅拌机，搅和器
AGV（Automated Guided Vehicle）自动引导运输车辆
AGV（automated guided vehicle）//n. 自动引导运输工具
air cushion plate 气垫板
air pipe 气管
air vent 排气道，排气槽
air-bend die 自由弯曲模
air-cushion eject-rod 气垫顶杆
alcohol container 酒精容器
algebraic 代数的，代数学的
align 对准，排成直线，排列
alignment 队列，对准，同轴度
Allen-Bradley A-B 公司
allocate 分派，分配
allowance （加工）余量，容差
alloy steel 合金钢
alloy tool steel 合金工具钢
alphanumeric 文字数学的，字母与数学结构的
alter 改变
alteration 改变，变更
always 总是，始终
AMMA　Acrylonitrile/methyl Methacrylate 丙烯腈/甲基丙烯酸甲酯共聚物
analog-mode device 类模器
analogy 类似，相似，模拟
analyze 分析，分解
angle ejector rod 斜顶杆
angle from pin 斜顶

angle pin 斜导边
angular 有角的
angular misalignment 角位移，角偏差
angular surface 斜面
animation 活泼，有生气，特技，动画，机构模式
anisotropic 各向异性的
annealing 退火
annealing 退火（的）
annotation 注解，评注，注释
anode （电）阳极，正极
anti-clockwise（anticlockwise）逆时针
antiflowback valve 逆流阀
anti-impact strength 抗冲击强度
apart from 除……以外
appearance 出现，显露，外貌，外表
applicable 可适用的，合适的
application 使用，应用
application form for purchase 请购单
appreciable 可估计的，明显的，可观的
approach 接近，逼近，方法，步骤，途径，通道
approval examine and verify 审核
approximately 近似的，合适的
apron 拖板箱
APT（Automatically Programmed Tools）自动编程语言
arbor 刀杆，心轴
arc welding 电弧焊
area 范围，面积
argon welding 氩焊
arm 臂
arm radial 摇臂
ARP　Aromatic polyester 聚芳香酯
arrangement 排序，安排，布置
AS　Acrylonitrile-styrene resin 丙烯腈-苯乙烯树脂
as a result 结果，因此
as a rule 通常，照例

as follows 如下
as result of ……结果，由于
as well 同样，也
as well as 以及，和……一样
ASA Acrylonitrile-styrene-acrylate 丙烯腈/苯乙烯/丙烯酸酯共聚物
asbestos 石棉
assembing die 复合冲模，装配用模具
assemble 安装，装配，组件
assembly line 组装线
assembly 集合，装置，集会，集结总成组件
assistant manager 助理
associate 联合，结合
associativity 集合性，缔合性
assume 呈现，表现为……形式
at last 终于
at least 至少
at right angles to 与……成直角
at the rate of 按……速率（比率）
attach 附上，加上
attach to 附在……上
attachment 附件机构，装置
attempt to 试图，企图
attraction 吸引，引力
audit 审计，稽核，检查
automate 使自动化
automated guided vehicle 自动搬运车辆，自动搬运交通工具
automatic 自动的，机械的
automatic screwdriver 电动起子
automatic single-cycle operation 自动单循环工作，自动单循环操作
automatic tool changers 自动换刀架
automatically 自动地
automation 自动化
automobile [美]汽车
automotive 自动的，汽车的
automotive industry 汽车工业

auxiliary function 辅助功能
auxiliary 辅助（的），副（的），备用（的），附属（的），补充（的），补助（的）附
availability 存在，具备，有效，利用率
available 可用到的，可利用的，有用的
axes 轴（复数）
axial 轴的，轴向的
axial thrust 轴向推力
axially 轴向地，与轴平行地
axis 轴
axis 轴，轴线，坐标轴，中心线
axisymmetric//n. 轴对称

B

back and forth 来回，往复
Backflowing 逆转的，回流的
background 背景，后台
backing plate 垫板，背板，支承板
backlash 间隙，轮齿隙
back-stroke ring 回程圈，回程环
baffle plate 挡块
baffle 导流块
balance 平衡
ball catch 波子弹弓
ball screw 滚珠丝杠
ballistic-particle manufacturing 喷射颗粒制造
ball-nut 球状螺母
band heater 环带状的电热器
band 带子，镶边，波段，队，联合，结合
band-aid 创可贴
bar 条钢，型钢
barcode scanner 条码扫描器
barcode 条码
barrel 圆筒，桶，圆柱体，燃烧室；把……装桶
base 床身
basket 蝴蝶篮
batch manufacturing 批量生产，批量制造

New Words

batch production 批量生产
batch 一炉，一批
be adaptable to 合适的，适合的
be comprised of 包含……，由……组成，由……构成，包括在……内
be parallel to 与……平行的，平行于……
be perpendicular to 与……垂直的，垂直于……
be suitable for 适用于……，适合于……
be attached to 连接于，固定于
be concerned with 与……有关系
be divided into 分成……
be grouped into 分类的
be known as 称为，叫做
be provided with 装有，备有
be put in storage 入库
be recognized as 被认为是
be subjected to ……易受到……
bead 珠子，水珠
bearing 轴承
bed 床，床身
bed die 下模，底模
bedway 热身导航
behavior 举止，行为，性质，状态，性能，习性
belong to 属于
belt drive 皮带传动
bend 弯曲
bend ability 可弯性
bend arc 弯曲弧
bend length 弯曲中性层弧长
bending angle（line） 弯曲角（线）
bending block 折刀
bending brake（bending machine）弯扳机，拆弯机
bending die 弯曲模
bending equipment for plastics 塑料折弯设备

bending fatigue 弯曲疲劳
bending fixture 弯曲夹具
bending force 弯曲力
bending moment diagram 弯钜图
bending operation 弯曲工序
bending press 弯曲压力机，弯曲机
bending radius 弯曲半径
bending strength 抗弯强度
bending strength 抗弯强度
bending test（stress，work）弯曲试验（应力，加工）
bending with sizing 校正弯曲
bending 弯曲
Bendix 邦迪克斯公司
benefit 利益，好处
bent pilot 弯曲导正
best-known 众所周知的，典型的
bevel 斜角，斜面，斜角的，
bevel gear 伞齿轮
bezel panel 面板
bidirectional 双向的
bilateral 双向的
binder 黏结剂
bismuth 铋
blade 刀刃，刀口，叶片
blank 毛坯，坯料
blank holder 坯料夹紧器
blank length of bend 弯曲件展开长度
blanking clearance，die clearance 冲裁间隙
blanking die 冲裁模
blanking force 冲裁力
blister 水泡
blow 打击，冲击
blow mold 吹塑模
blow molding 吹塑成形
board 看板
body 体，钻体
body clearance 钻体间隙
bolster plate 工作台

163

bolster size 工作台板
bolster 垫枕，垫木，垫枕状的支持物，支撑，垫，支持
bolt 螺栓，螺钉，用螺栓固定
BOM (bill of material) 材料清单，材料表
Boolean algebra 布尔代数
bore 镗孔
boring 镗孔
boring bar diameter 镗杆直径
boring 镗削
bottom clamp plate 底模侧固定板，动座板
bottom stop 下死点
branch 分支，部门
branch runner 分流道，次流道
braze 铜焊，硬钎焊
brazing 铜焊（接），硬钎焊
break down 破坏，毁坏，击穿
breaking．(be) broke, (be) cracked 断裂
breakthrough 突破，破损
brine 盐水
Brinell test 布氏硬度试验
bring to bear on 施加（影响，压力）
brittleness 脆性，脆度，易碎性
broad 宽的
broad 宽的，阔的，广泛的，显著的，主要的，宽阔部分
broadly 概括地，大致地
broadly speaking 概括地讲
brown marks/burnt streak/silver streaks 焦痕，条纹，银痕，条痕
buckling 扣住
buffer 缓冲器
buffle 隔片
built-in software 内置软件
bulging 胀形，鼓凸，撑压内形
bulk 大小，体积，大批，散装
bullet 子弹
burnish 磨光，擦亮（金属）

burr 毛刺，毛边，轴环，套环，垫圈，去毛刺
bush（机械）衬套，轴瓦，加衬套于……上
bushing 轴套，衬套
bushing block 衬套
bushing/guide bushing 边司/导套
buzzle 蜂鸣器
by comparison 相比较
by means of 依靠
by no means 决不
by steps 逐步地
by-pass 旁边，迂回

C

CA Cellulose acetate 醋酸纤维塑料
CAB Cellulose acetate butyrate 醋酸-丁酸纤维素塑料
cabinet （有抽屉或格子的）橱柜
CAD （Compter-Aided Design）计算机辅助设计
cadmium 镉
CAE （Compter- Aided Engineering）计算机辅助工程
cage 笼，笼盒，隔离罩
calculate 计算
calendaring 延压成形
cam 凸轮，偏心轮
CAM （Compter-Aided Manufacture）计算机辅助制造
camouflage 伪装，掩饰，掩盖
cancellation 取消
cantilever 悬臂，交叉支架角撑架
CAP Cellulose acetate propionate 醋酸-丙酸纤维素
CAP （Compter-Aided production）计算机辅助生产
capability （实际）能力，性能，容量，接受力

New Words

capable 有能力的
capable of 能……的，可以……的
capacity 容量，能力
CAPP（Compter-Aided Processing Planning）计算机辅助工艺设计
CAQA（Compter-Aided Quality Assurance）计算机辅助质量保证（系统）
carbide 碳化物
carbide alloy steel 硬质合金钢
carbon steel 碳素钢
carbon tool steel 碳素工具钢
carburize 渗碳
carburizing 增碳剂，渗碳剂
care 管理，照顾
career card 履历卡
careful 小心的，仔细的
carefully 仔细地
carriage 溜板，拖板
carriage 拖板，客车，运费，车架
carton box 纸箱
carton 纸箱
cartridge heater 发热管
cascade 小瀑布，喷流，成瀑布落下
铸件，铸造
case iron 铸铁
case iron permanent mould die 铸铁持久注射模
case steel 铸钢
case-hardened steel 表面淬火钢，表面硬化钢，表面渗碳钢
cast 铸件，铸造
cast iron 铸铁
casting 铸造，铸件
casual 偶然的，不经意的，临时的
category 种类，类别，范畴，类型
cathode 阴极，负极
cathode-ray tube(CRT) 阴极射线管
cavity direct cut on A-plate，core direct cut on B-plate 模胚原身出料位
cavity insert 上内模
cavity plate（block）凹模，凹模板
cavity retainer plate 定模模板
cavity sinking EDM machines 型腔电火花加工机床
cavity 型腔，母模
CE Cellulose plastics 通用纤维素塑料
celluloid 塞，明胶 adj. 细胞状的
cellulose nitrate 硝酸纤维
cellulose 纤维素
cementite 碳化铁
center 中心，顶尖，定中心，
center height from floor 主轴中心离地高度
center line 中心线
center of die，center of load 压力中心
center support ring 中心支撑环
centerless grinder 无心磨床
centrifugal 离心的
centrifugal force 离心力
centrifuge 离心机，分离机
CF Cresol-formaldehyde 甲酚-甲醛树脂
CG（computer graphics）计算机图形，计算机图形学
chain 链条，链条槽
chamber 室，盒
change-over 转换，改变，更换，倒转
channel 通道，沟槽
character die 字模
character recognition 字符识别
characteristic 特性，特征
characterize 表现……的特色
charcoal 木炭
charge 费用
chassis 基座
checklist 核对清单，检查清单
chilled 已冷的，冷硬了的，冷冻的
chip 碎屑，切屑，细碎物
chisel 凿，錾

chop 砍，排骨，切断，品种
chromium 铬
chromium steel 铬钢
chromium 铬
chuck 卡盘，夹住
CIM（computer intergrated manufacturing）计算机集成制造
circle 圆周，圆，环绕
circuit diagram 电路图
circularity 圆度
circumferential 周向的，圆周的
clamp 夹子，夹具，夹钳，压板，压铁
clamp plate（ring）压板（夹紧环）
clamping force（element, device, piston）夹紧力（件，装置，活塞），锁模力
clamping mechanism 锁模机构
clamping stroke 开模行程
clamping system 夹紧系统，锁模系统
clamping tonnage 锁模力
classification 整理
classified as 分为
classify…into… 把……分类为
clay 泥土，黏土
cleaning cloth 抹布
cleanness 清扫
clearance between punch and die 凹凸模间隙
clearance 余地，间距，切口，清除
climb 攀登，爬
climb milling 顺铣
cling 粘紧，附着，紧贴
clockwise 顺时针的，顺时针，顺时针方向的
close 关闭
close to each other 相互靠近
closed-loop control 闭环控制
closed-loop system 闭环系统
closed-type single-action crank press 闭式单动（曲柄）压力机

closely 接近地，紧密地
close-tolerance 紧公差，精密工差
clutch 离合器
CMC Carboxymethyl cellulose 羧甲基纤维素
CN Cellulose nitrate 硝酸纤维素
CNC machining center 数控加工中心
CNC system 数控系统
coarse 粗糙的，使粗，粗化
coat 涂上
coating 涂层，覆盖层
coefficient 系数
coining die 压印模
coining 冲制，压边，整形，模压，压花
co-injection 同时注射，并列注射，联动注射
co-injection molding 同时注射成形，共同注射成形
cold chisel 冷錾
cold slag 冷料渣
cold work die（tool）steel 冷作模具（工具）钢
cold-runner mold 冷流道模具
cold-runner three-plate mold 冷流道三块板模具
cold-runner two-plate mold 冷流道二块板模具
cold-slug well 冷料穴，冷料井
collapse 倒塌，瓦解
collar 领，环，法兰盘
collision 碰撞，冲突
colorable resin 可着色树脂
column 立柱，圆柱，专栏，柱状物
combination 组合，组合物
combination injection 联动注射,合并注射
combine with 与……相结合
come about 发生，出现
come in contact with 同……接触
command 命令

New Words

commercial 商业的，工业用的，工厂的，大批生产的
common equipment 常用设备
comparatively 比较地，相当地
comparison 比较
compatible 相容的，兼容的，协调的
compensate 补偿，偿还
compensation 补偿，赔偿
compensative function 补偿功能
competitive 竞争的
compile 编辑，汇编
complete 完成，使完善
complex 复合体
complexity 复杂性，复杂的事物，复杂性
component 元件，组件
component drawing 零件图
composite die，combined die 组合模，拼合模
compound blank and pierce dies 落料冲孔模
compound die 复合模
compound piercing die 复合冲孔模
compound 合成的，复合的，复合，组合
compre sion molding 压缩成型
comprehensive//adj. 全面的，广泛的，综合的
compressing mold 压缩模（也叫压制模）
compressing molding 压缩成形
compressing stage 压缩段
compression 压紧，浓缩，压缩，压力，压榨
compromise 折中
computational 计算的，计算机的
computer aided design(CAD)
computer aided manufacturing(CAM)
computer graphics 计算机绘图
computer integrated manufacturing system (CIMS)
computer numerical control(CNC)
computer workstation 计算机工作站
computer-aided die design 计算机辅助模具设计
computer-aided NC programming 计算机辅助数控编程
computer-aided programming 计算机辅助编程
computer-based drafting package 基于计算机工程图软件包
computer-based visualization technology 基于计算机的可视化技术
computer-generated prototype 计算机生成的原型
concave 凹的
concentricity 同轴度
conceptual 概念的，构想的
conceptual design 概念设计
conceptual model 概念化造型，概念化建模
conceptualization 化为概念，概念化
concurrent 同时发生的事件
conduct 实施，进行
conductivity 传导性，传导
configuration 构造，结构，配置
conform 使一致，使……符合，
confusion 混乱
conical 圆锥（体、形）的，圆锥形的
connector plug 插头
connector socket 插座
consensus 一致同意，多数人的意见
consequence 结果，推理，推论，因果关系
consequently. 从而
conservation 清洁
considerable 相当大（或多）的，
considerably 相当大地
consideration 考虑

consist of 包含，包括，由……组成
consistency 结合，坚固性，浓度，密度，一致性，连贯性
console 安慰，慰问，控制台，扎架
constituent 组成的，构成的
constitute 构成
consume，consumption 消耗
consumption 消费
contact 接触，触点，联系
continuous 连续的，持续的
continuously 连续不断地
continuous-path tool movements 刀具的连续轨迹运动
contour 轮廓，仿形
contoured surface 仿形面，围线曲面
contraction 收缩，缩写式，紧缩
control 控制
control board 控制板
control hierarchy 控制层次，控制层面
control panel 控制面板
control system 控制系统
conunterbore 扩孔，沉孔
conventional 一般的，常规的
conventional machining 常规加工
conventional milling 逆铣
conventional 惯例的，常规的，习俗的
conversion 变换，转化
convert 使转变
convert…into 把……变成
conveyer belt 输送带
conveyer 流水线物料板
conveyor 传送装置，传送带
coolant 冷却液
coolant tank capacity 冷却水箱容量
coolant 冷冻
cooling channel 冷却流道
coordinate 坐标
coordinate system 坐标系，坐标系统
copper 铜

copper electrode 铜公（电极）
core insert 下内模
core locking wedge 型芯锁定楔块
core retainer plate 动模模板
core 核心
core/cavity inter-lock 内模管位
core-pulling force 抽芯力
correspond 符合，相应
corresponding 相应的，符合的，一致的
corresponding 相应的，通信的，相当的，对应的
correspondingly 相对地，相比地
corrosion 腐蚀
corrosion-resistant steel 耐腐蚀钢
corrosive 侵蚀的
cosmetic defect 外观不良
cosmetic inspect 外观检查
cost 成本
counterbore 铁锈，锈
counterboring 平底锪钻
counterclockwise 反时针方向的
countersinking 尖底锪钻
coupling 联轴节，联轴器
coutout 轮廓，外形，等高线
cover plate 盖板
CP Cellulose Propionate 丙酸纤维素
CPE Chlorinated Polyethylene 氯化聚乙烯
CPVC Chlorinated Poly（vinyl chloride）氯化聚氯乙
crack 裂缝，断裂
cram 填塞，压碎
crank 曲轴，曲柄，手柄，弯曲
crater 弹坑
creep 爬，蔓延
crevice 裂缝
criterion 标准，准据，规范
critical defect 极严重缺陷
critical dimension 重要尺寸

critical range 临界点
critical range 临界范围
critical temperature 临界温度
critical 评论的，签定的，批评的，危急的，关键的，临界的
cropping 切断
cross slide 横向拖板
crosshatching 双向影线，十字口
cross-linking 交叉连接的
cross-sectional area 横截面积
cross-sectional view 横截面剖视图
cross-sectonal 横截面的
CRT Character Display 屏幕特征显示
CRT（Cathode-ray Tube）显像管，阴极射线管
CS Casein 酪蛋白
CTA Cellulose triacetate 三醋酸纤维素
cubic 立方体的，立方的
culture 教养
cumbersome 繁重的，麻烦的，笨重的
cure 治疗，纠正
curing temperature 固化温度
curing 固化
curl 使卷曲，使卷边
curling die 卷曲模
current 电流
cursor control 光标控制
cursor 指针，光标
curve 曲线，绘曲线
curvilinear 曲线的
cushion 缓冲器
customarily 通常，习惯上
customary 习惯的，通常的
customize 定制，用户上
custom-made 专门制造的，按照要求制造的
cutter 刀具
cutter compensation 刀具补正
cutter grinder 工具磨床

cutter saddle 刀架
cutting die，blanking die 冲裁模
cutting edge 切削刃
cutting traverse 切削进给
cutting-off die 切断模
cyanide 氰化物
cycle functon 循环功能
cylinder bushing 汽缸套
cylinder 汽缸套，圆筒，圆柱体，柱面
cylindrical 圆柱（体，形）的
cylindrical pin 圆柱销

D

D.C.motor 直流电机
data processing unit(DPU) 数据处理单元
database 数据库，资料库
database 数据库，资料库
DC servo motor 直流伺服电动机
DC（Direct Current） 直流电流
dead center 尾架顶尖，死点，止点
deal 量，分配
deal with 处理
debug 调试
debugging 调试
deburring 去毛刺
deburring punch 压毛边冲子
decimal 十进的，小数的
decimal point 小数点
declining 倾斜的，衰退中的
decode 解码，译码
decomposition 分解，腐烂，还原，分离，风化
decompression 减压，解压，分解
deep 深的
deep-drawing die 拉深模，深拉模
defect 过失，缺点，缺陷
defective product box 不良品箱
defective product label 不良标签
defective products n ot up-to-grade products 不良品

defective to staking 铆合不良
defective upsiding down 抽芽不良
defective 有缺陷的，欠缺的
deficient manufacturing procedure 制程不良
deficient purchase 来料不良
deflection 偏转，挠曲
deflection coil 偏转线圈，偏转绕组
deformation 变形，形变，扭曲，失真
degassing stage 排气段
degree 度，程度
degree of accuracy 精度，精确度
degree of freedom 自由度
delete 删除
deliberate 深思熟悉的，有准备的，故意的
deliberately 故意地
delivery deadline 交货期
delivery，to deliver 交货
demand and supply 需求
demonstrate 证明，论证，示范
demould 脱模
denote 指示，表示
dent 压痕
department director 部长
depend 依靠，依赖，取决于
deposition 沉积物，沉积作用
depression 消沉，底气压，底压，下降
deputy manager =vice manager 副经理
deputy section supervisor =vice section superisor 副课长
derive 派生，起源
derived from 起源于，来源于
describe 描写，记述
design 设计
design modification 设计变化
designation 指定，名称，指示
designer 设计师
designing，to design 设计
detect 发觉，察觉

deterioration//n. 变坏，退化，失效
detriment 损害，损害物
deviation 偏差
devise 设计，发明
diagnoses 诊断，评价，调查分析
diagnostic 诊断的，特征的
diaphragm gate 隔膜浇口
die 冲模，锻模
die and mold 模具，冲模
die block 模块
die block steel 模具钢
die change 换模
die for special purpose 专用模具
die insert 模具型芯，模具嵌件，模具镶块
die insert waterway 模具嵌入式流道
die life 冲模寿命
die lifter 举模器
die locker 锁模器
die material 模具材料
die plate，front board 模板
die repair 模修
die shut height 模具闭合高度
die space adjustment 装模高度调节量
die spring 弹弓
die worker 模工
die 模具
die-casting die 压力铸造模具，压铸模
die-casting machine 压铸机，压力铸造机
dielectric characteristics 电绝缘性能，介电性能
dielectric 电介质，绝缘体，绝缘材料
differentiation 绝缘材料，绝缘体
digitize 区别，差别
digitizer 将资料数字化，数字转换器
dimension 尺寸，标尺寸
dimension/size is little bigger 尺寸偏大（小）
dimensional accuracy 尺寸精度
dimensional 空间的

New Words

direct current=D.C 直流电
direct current=D.C 直流电
direct sprue gate 唧嘴直流
direction 方向，指导
disadvantageous//adj. 不利的，缺点的，不足的
discharge 放电，放出
discoloration 异色
discoloration 变色，污点，褪色，脱色
discover 发现
discrepancy 相差，差异，矛盾，配置
discrete 离散的
dish gate 因盘形浇口
dishwasher 洗碗机，洗碗的人
dismantle 分解（机器），拆开
dissimilar 不同的，相异的
distance between centers 顶尖最大夹持长度
distance between uprights 立柱间距离
distinguish 区别，辨别
distortion 扭曲，变形，失真
distribution 分配，分发，配给物，分类，发行
Do not use（core/cavity）insert 不准用镶件
documentation 文件，记录，提供文件
dog 挡抉，鸡心夹头
dome 圆盖，圆屋顶
dominant 有统治能力，占优势，显性的
DOS（Disk Operating System）磁盘操作系统
dosposal 不连续的
double 两倍的，使加倍，加倍
double V-die 双 V 形弯曲模
double-acting 往复式的
double-acting press 双动式压力机
double-acting traverse cylinder 双动横向汽缸
double-housing planer 双立柱（臂）龙门刨床

double-row radial bearing 双列径向轴承
dovetail 与……吻合
dowel pin 管钉，销子
dowel 木钉，暗销，定位桩
down milling 顺铣
downward 向下的
draft 脱模斜度，拔模斜度，草稿，草案，草图
drafting 制图
drastic 强有力的，激烈的，猛烈的
draw 拉伸
drawing 绘图，拉长，拉深，拉拔
drawing die 拉深模
drawing machine 拉拔机
drawing numbers 拉深次数
drawing ratio 拉深比（系数，力，速度）
drill 钻床，钻孔
drill press 钻床
drilling 钻削
driven element 被动元件
driver 发动机，主动轮
driving element 主动元件
drop forging 模锻，落锤锻
droplet 小滴，液滴
dry run 空运行
dual 双的，二重的，双重
ductility 延（展）性，韧性，延展性，延伸度，可锻性
due to 由于
dull 变钝，弄钝，钝的
durable 持久的，耐用的
duty 责任，功率，税
dwell time 保压压力
dye 染料，染色
dynamics 动力学

E

early return bar 复位键、提前回杆
easily damaged parts 易损件
EC　Ethyl Cellulose 乙烷纤维素

eccentric 偏心的，中心不对称的
economically 经济地，节约地
edge 刀口，刀刃，边缘
edge gate 大水口，侧缘浇口
EDM grinder 电火花加工磨床
EDM（Electrical Discharge Machines）电火花加工机床
EDP（Electronic Data Processing）电子数据处理
education and training 教育与训练
effective holding pressure 有效保持压力
effectiveness//n. 有效性，有效，效力
effector 受动器，执行件
efficient 有效率的，能干的
either…or 或……或，不论……
eject pin 顶出针
eject rod（bar）（成型机）顶业捧
eject 弹射，喷射
ejector cylinder 顶出汽缸
ejector force 顶出力
ejector guide bush 中托司
ejector guide pin 中托边
ejector guides bush 推板推套
ejector pin 顶针，司筒针
ejector pin retaining plate 推杆固定板
ejector plate 顶出板，推出板，顶杆垫板
ejector retainner plate 顶针板，顶销固定板
ejector return pin 顶出复位杆
ejector sleeve 司筒
ejector stroke 顶出行程，托模行程
elaborate 精心制作的，精细
elastic 弹性的
elastic deflection 弹性（挠曲）变形
elastic limit 弹性极限
elastic recovery 弹性恢复，回弹
elastic stress 弹性应力
elasticity 弹力，弹性
elastomer 弹性体，人造橡胶
elbow 肘
electric heater band 带电式加热圈
electric screw driver 电动起子
electrical 电的，电力的
electrical discharge machine 电火花加工机床
（EDM）电火花加工
electrical insulating properties 电绝缘性能，介电性能
electrical sparkle 电火花
electrically 用电，电学上
electricity 电，电学
electrochemical 电气化学的
electrochemical die-sinking machine 电解型腔加工机
electrochemical grinding 电化学磨削
electrochemical machine（ECM）电解加工机床，电化学加工机床 电解加工，电化学加工
electrochemical//adj. 电气化学的
electrode 电极
electro-discharge cutting machine 电火花切割机
electro-discharge machine 电火花加工机床
electro-discharge machine tool 电火花加工机床
electrolyte 电解质，电解液，电离质
electron beam cutting machine 电子束切割机
electron beam machine tool 电子束加工床
electron beam machining（EBM）电子束加工
electron-beam cutting installation 电子束切割装置
elevate 升高，提升，举起，使升高
eliminate 排除，消除，除去

New Words

elongate 拉长，伸长
EMA Ethylene/methacrylic acid 乙烯/甲基丙烯酸共聚物
embedded lump 镶块
emboss 使浮雕出来
emergency 落出，发生，出现，突然事件，紧急事件
EMI gasket 导电条
emphasize 强调，着重，重点
empirical 经验的，实验的
employ 使用，采用
encapsulate 包装，封装，压缩，形成胶囊
encapsulation molding 低压封装成型
encircle 环绕，围绕，包围
encode 编码，编码器
encoder 编码员，编码器
encompass 包含或包括某事物
encompass 包围，环绕，包含或包括某事物
energize 加强，给……电压，给予精力
energy 能力，[物]能量
energy balance 能量平衡
energy-absorbing collision 能量吸收碰撞
engage 接合，啮合
engine 发动机
engine lathe 普通车床
engineering consulting office 工程咨询机构
engineering, project difficulty 工程瓶颈
engineering 工程
engraving 刻印
enhance 提高，提强
enhanced 提高的，提强的，放大的
enlarge 扩大
enlargement 放大，扩大
enough 足够的，充足，足够地
entail 使必要，使蒙受
enterprise plan = enterprise expansion projects 企划

entire 全体的，完全的
entirely 完全地，彻底地
entrap 收集
environmental 周围的，环境的
envision 想象，预想
EP Epoxy，epoxide 环氧树脂
EPD Ethylene-propylene-diene 乙烯-丙烯-二烯三元共聚物
EPM Ethylene-propylene polymer 乙烯-丙烯共聚物
epoxy resin 环氧树脂
epoxy 环氧，环氧树脂
EPS Expanded polystyrene 发泡聚苯乙烯
equation 相等，方程式，等式
equator 赤道
equilibrium 平衡，均衡，保持平衡的能力
equipment 装备，设备，装置
eracuretun 顶板
erase 擦掉
erosion 腐蚀，侵蚀
erratic 不稳定的
error 错误，过失，误差
essential 主要的，根本的，基础
establish 确定，建立
ETFE Ethylene-tetrafluoroethylene 乙烯-四氟乙烯共聚物
eutectoid//n. 类低共熔体的，共折体
EVA Ethylene/vinyl acetate 乙烯-醋酸乙烯共聚物
EVAL Ethylene-vinyl alcohol 乙烯-乙烯醇共聚物
evaluation 评估，计算，估价
even cooling 均匀冷却
even if 即使
even number 偶数
even though 即使
evenly 均匀地，平坦地

eventual 最后的，结局的，可能发生的，万一的
eventually 最后，终于，最后地，终于地
evident 明显的，显然的
evolve 进化，演变
examine 检查，调查，考试
exceptionally 格外地，异常地
excess 过度，超过，过度的
excessive defects 过多的缺陷
excessive 过分的，过多的，额外的
excesssive gap 间隙过大
exclude 拒绝接纳，排斥
exclusively 排外地，专有地，排除在外地
execute 执行，实行，完成
execution 实行，完成，执行，制作
exert 发挥，施加，远行，行使
expanding die 胀形模，扩管模
experimental 试验（用）的
explain 解释，说明
explosion 爆发，发出，爆炸
exponent 指数，倡导者
exposed metal/bare metal 金属裸露
extend 扩充，延伸
exterior 外面的
exterior cylinder 圆柱体外表面
external grinder 外圆磨床
extraordinary 非常的，特别的，非凡的
extremely 极端地，非常地
extruder 压出机，挤压机
extruding 挤压出，挤压成，突出，伸出，逐出
extrusion die 挤压模具
extrusion 挤出，挤压，挤压成型，压出品，推出

F

fabric 纤维品，构造
Fabricate 制作，构成，伪造，虚构
factory 工厂
factory director 厂长
Fagor 发格公司
Fahrenheit 华氏温度计的，
Failure 故障，断裂
fall into 落入，开始，把……分成，分类
familiar 熟悉的
familiar with 熟悉……的
familiarity 熟悉，通晓，亲密，精通
fan gate 扇形浇口
fan-shaped block 扇形块
Fanuc 法纳科公司
faraday's law 法拉第定律
fasten （使）固定，加固，连接，锁紧（螺丝）
FDD（floppy disk drive）磁盘驱动器
feasibility 可行性，可能性
feature change 特性变更
feature-based solid modeling technology 特征实体建模技术
feed 进给，送给
feed hopper 供料仓，给料仓
feed rate override 进给倍率开关
feed（gating, runner）system 浇注系统
feedback encoders 反馈编码器
feedback unit 反馈单元
feedback 反馈，回馈
feeder 送料机
feeding is not in place 送料不到位
feeding stage 加料段，喂料段
FEP Perfluoro（ethylene-propylene）全氟（乙烯-丙烯）塑料
ferrite 铁酸盐，铁素体
FF Furan formaldehyde 呋喃甲醛
fiber-reinforced 增强纤维的，加强纤维的
fibric belt 纤维带
field strength 电场强度
figure file, chart file 图档
filament 细丝，灯丝
filings 铁削
filled resin 充填树脂

New Words

filler 填料，装饰物，过滤器，滤纸 过滤
filling velocity 保压时间
film gate 薄膜浇口
final inspection 终检
fine 好的，优良的，精细的
finish 精加工，粗糙度，表面抛光，
finished piece 完成件
finished product 成品
finite-element mesher 有限元网格划分器（模块）
finite-element method（FEM）有限元方法
fire resistance 耐火性
fit 装配，配合
fit together 组装在一起
fix 解决，调整，修理
fixed clamping plate 固定侧固定板，定座板
fixed compression mold 固定式压缩模
fixed plate 固定板
fixed retainer plate 定模模板
fixed stock stop pin 固定挡料销
fixed transfer mold 固定式传递模
fixed-column 固定立柱
fixture offset code 固定偏置代码，固定偏置编码
fixture offset method 固定偏置法
fixture 夹具，夹紧装置，型架，固定
flagship 旗舰
flame cutters 切割机，火焰切割机
flange 边缘，轮缘，凸缘 vt. 给……装凸缘
flash gate 闸门浇口
flash mold 溢流式模具
flash ridges 飞边值，飞边凸峰值
flash 闪光，一瞬间，溢出式数模，溢料，毛刺，突然现象的
flash-type mold 溢出式压缩模，飞出式压缩模
flatness 平面度

flexibility 机动性，挠性，弹性，适应性，机动性
flexible 灵活的，能变形的
flexible manufacturing cell 柔性制造单元，柔性生产单元
flexible manufacturing system 柔性制造系统
flexible modes 柔性模式，柔性方式
floor to table surface 地面至工作台面高度
floppy diskette 软盘
floppy 软的，松软的
flow board 流水板
flow chart 流程图
flow mark 流痕
fluid 流体，液体，
flushing 冲洗，洗净
flute 凹槽，螺旋槽
flux 助熔剂，焊剂
flywheel 飞轮，惯性轮，整速轮
flywheel-type screw press 飞轮式螺旋压力机
FMS（Flexible Manufacturing System）柔性制造系统
foam molding 发泡（注射）成形
foam 泡沫，水沫，泡沫材料 vi. 起泡沫 使起泡沫
fold of pakaging belt 打包带折皱
folded block 折弯块
follower 从动件，随动件
fomaldehyde 甲醛
font file 字体文件
footstock 尾架，尾座
for instance 例如
forefinger 食指
forge 锻工车间，锻造
forged weld 锻焊
forging 锻造，锻件，锻制的
forklift 铲车
form grinding 成形磨削

175

format a diskette 磁盘格式化
formation 成形，形成，构成
formed steel 型钢
former 从前的，以前的
formerly 以前，过去
forming die 成形模
forming technology 成形工艺，成形技术
formula 公式，规则
found application in 在……方面得到使用
four-crank press 四曲柄压力机
four-jaw(independent)chuck 四爪（分动）卡盘
four-point press 四点式压力机
fracture 断口，断面，断裂 断裂，拆断
frame buffer 帧缓存器
frame 框架，机架
framework 构架，框架
frequency 频率，频繁
frequent 时常发生的，频繁的
frequently 常常，屡次
friction screw press 摩擦传动螺旋压力机
frictional 摩擦的
front plate 前板
full automatic operation 全自动工作，全自动操作
function 作用，[数]函数
fundamental 基本原则，基本原理
fundamentally （从）根本上
furnish 供应，提供，装备，布置
furthermore 此外，而且
fuse 熔合
fuse machine 热熔机
fuse together 熔合
fused-deposition molding 熔丝沉积成形
G
gap 间隙，裂口，缺口，开口，使成缺口
gap-frame press 开发压力机
garbage bag 垃圾袋
garbage can 垃圾箱

garbage container 灰箕
garner 石榴石，深红色
gas-assisted injection molding 气体辅助注射成型
gaseous 气体的，气态的
gate location 进入位
gate size 水口大小
gate type 水口形式
gate 门，浇口，内浇口
gauge 量规
gauge block 量块
GE 通用电气公司
gear 齿轮
gear mechanism 齿轮机构
gearing 传动装置，齿轮装置
general manager 总经理
general-purpose 通用的，多方面的，多种用途的
generative system 生成系统
generative 生成的，产生的
generic 属的，类的，一般的，普通的
geographical 地理学的，地理的
geometric 几何的，几何学的
geometry 几何形状，几何学
geometry modeler 几何图形建模器
geometry molding 几何建模
germanium 锗
gigantic 巨大的，巨人般的
gland bushing 密封管套
glass fiber 玻璃纤维
globular 球状的，有小球的，世界范围内的
glove（s）手套
glove（s）with exposed fingers 割手套
glue 胶水，胶，使结晶，使成细粒 形成细粒
good stability，stable 稳定性好
gradient//adj. 倾斜的，梯度，倾斜度，坡度

New Words

graduate 刻度
grain 谷物，晶粒，粒度，使结晶，使成细粒　vi. 形成粒状
granular 粒状的
granule 小粒，颗粒，细粒
graphic 绘画似的，图简的
graphically 绘成图画似地，绘画地
graphics 制图学，图形
graphics scanner 图形扫描仪
graphics 制图法，制图学，图表算法，图形
graphite 石墨
gravitational 重力的
gravity 地心引力，重力，重要性，严重性
gray-iron 灰铸铁
grease/oil stains 油污
grinder 磨床
grinding machine 磨床
groove 沟，槽，开槽
groove punch 压线冲子
gross weight 毛重
groundwork 地基，基础，根基
group leader/supervisor 组长
group technology 成组技术
group technology(GT) 成组技术
guarantee//n. 保证书，保证，担保，抵押品，保证，担保
guide bushings 导套
guide pad 导料块
guide pillar 导柱，导杆
guide pin 导向销，导杆
guide pin bushing 导套
guide plate 定位板
guide post 导柱
guide stock plate 挡料板
guideline//n. 指导方针，指导性文件

H

hallmark//n. 检验，品质证明，特点
在……上盖上纯度检验印记
hallmark//n. 检验，品质证明，特点
在……上盖上纯度检验印记
handle 柄，把手
handle mold 手持式模具
handle 处理，操着的柄，把手
处理，管理，控制，操纵，操作
hard disk 硬盘
harden 淬火
harden 使硬，使变强，变硬，变冷酷
hardenability capacity 淬硬性（硬化能力）
hardenability curve 淬透性曲线
Hardened steel 淬硬钢
hardening 硬化，强化
hardness penetration diagram "U"形曲线
hardness profile 硬度分布（硬度梯度）
hardness 硬度，硬性，刚度
hardware 五金器具，硬件，部件
HDD（Hard Disk Drive）硬盘驱动器
HDPE High-density polyethylene plastics 高密度聚乙烯塑料
head of screwdriver 起子头
Headstock 主轴箱
heat resistance 耐热性
heat system 加热系统
heat time 加热时间
heat transmission 热传递，热传输
Heat treater 热处理工作者
Heat treatment 热处理
heat treatment cycle 热处理工艺周期
heat treatment furnace 热处理炉
heat treatment installation 热处理设备
heat treatment procedure 热处理规范
heater capacity 电热容量
heating curve 加热曲线
heating up time 升温时间
heat-resistant 耐热的
heat-resistant steel 耐热钢
heavy-duty 重型的，重载的
helical 螺旋形的

helical gear 螺旋线
hemming 折边，卷边
herringbone 人字形，箭尾形的
herringbone gear 人字齿轮
hierarchy 层次，层级
high alloy steel 高合金钢
high carbon steel 高碳钢
high production rate/high productivity 高生产率
high stiffness-to-weight ratio 高硬度-重量比
high temperature carburizing 高温渗碳
high temperature tempering 高温回火
high-frequency oscillation 高频振荡器
high-speed 高速的
high-speed punch press 高速冲床，高速冲压机床
high-speed steel 高速钢
hinder 阻碍，打扰
Hi-pot test of SPS 高源高压测试
HIPS High impact polystyrene 高冲聚苯乙烯
hob 滚刀，滚铣
hobbing 滚铣，滚齿，压制的，横压的
hobbing machine 滚齿机
hold 拿，握，支持，夹住，占据，持有，拥有
holding fixture 夹具
holding plate 支撑板
homogeneous 同类的，相似的，均匀的，单一的，单色的
honing 珩磨
hopper 漏斗，给料斗，记量器，料仓
horizontal 地平线的，水平的
horizontal machining center 卧式加工中心
horizontally 地平地，水平地
hot manifold 热流道板
hot runner，hot manifold 热流道，热浇道
hot sprue/cold runner 热嘴冷流道

hot work mold(tool)steels 热作模具(工具)钢
housing 套，壳，罩
hub 衬套
hydraulic 水力的，液压的
hydraulic clamp 液压夹紧
hydraulic fitting 液压装置
hydraulic handjack 油压板车
hydraulic lock 液压锁紧机构
hydraulic machine 油压机
hydraulic motor 液压电动机
hydraulic press 液压机，水压机
hydraulic reservoir 液压容器，液压储存器
hydraulic servo motor 液压伺服电动机，液压随动系统
hydraulic system 液压系统
hydraulic tubing 液压管道系统
hydraulic 液压的，液力的 流体力学的，液压机械的
hydroformer press 液压成形压力机
hypothetical 假设的，假定的

I

i.e.=that is 即，就是
ICG（Interactive Computer Graphics） 人机交互计算机图形
ideal 理想的
identical 同一的，相同的
identifying sheet list 标示单
identifying 识别，鉴别，把……看成一样，确定
illustrate 举例说明，图解，加插图于，阐明 //vi. 举例
image 图像，肖像，映像，典型 vt. 想象，反映，象征
immediate 直接的，立刻的
immediately 立即，马上，直接地
impact 冲击
impact strength 抗冲击强度
impinging 碰撞

New Words

implement 工具，器具，贯彻，实现，执行
implementation 执行，贯彻，履行
implemented 实行，执行
impression 凹腔，压痕，型腔
impressionable 易受影响的，敏感的，可塑的
in a similar way 同样
in addition 此外
in an efficient manner 用有效的方法，有效地
in contact with 与……接触
in contrast 相比之下，相反，可是
in effect 有效
in essence 本质上
in opposite direction 朝相反方向
in other instances 在其他情况下
in the direction of… 朝……方向
in the fifties 在五十年代
in the form of 以……形式
in the view of 由于，鉴于，考虑到
in this way 这样
inch 英寸
incline 倾斜，偏向 n. 斜坡，斜面
inclusion 杂质，包含，内含物
incoming stock 导正块（销），导入块（销）
incoming 引入的，导入的
inconsistency 不一致，不合理，矛盾
incorporate 合并，混合
incorporate 合并的，结社的，一体化的 结合，合并，插入，引入，编入，加入，
increment 增加，增量，增长
incremental mode 增量方式
incremental positioning 增量定位
incremental 增加的，增量的
indentation 压痕，压坑
indentation 缩排，呈锯齿状，缺口，印凹痕
independent 独立的，各自的，分开的，不受约束的

index（铣床）分度头，指针
indexing head 分度头
indexing table 分度工作台
indexing time 转位时间
indexing 标定指数，分度，指标
indicate 指示，读数
indiscriminately 随意地
individual 个人，个体，个别的，
individualize 个别地加以考虑
industrial alcohol 工业酒精
inertia 惯性，惯量
inexpensive 便宜的，不贵重的，廉价的
infeed 横向进给
infinitely-variable speed 无级变速
influence 影响，感化，势力，感应，影响，改变
infusible 难融化的，不融化的，不融化性的
ingot 铸锭，钢锭
ingredient 成分，因素
initiation 引发，开始
injection cylinder 注射缸，压射缸
injection mold 注射模，注塑模
injection mold for thermoplastics 热塑性塑料注射模
injection mold for thermosets 热固性塑料注射模
injection molding filling 注射成形充填
injection nozzle 注射喷嘴
injection plunger 注射柱塞
injection rate 注射速度，注射速率
injection speed 注射速度
injection stage 注射段，射出段
injection stroke 射出行程
injection 注射，注射剂，注入，射入，加压充满
injection/compression molding 注压成形
injection-molding machine 注射成形机
in-line 同轴的，嵌入的，内嵌的

179

in-line reciprocating screw system 同轴式往复螺旋系统，内嵌式往复螺旋系统
inner 内部的，在内的
inner parts inspect 内部检查
inner ram 内滑块
inner surface 内表面
innovative 创新的，革新的
innumerable 无数的
in-process gaging 在线检测
inquiry，search for 查寻
insert 镶件，入块（嵌入件），插入，嵌入，插入物，金属型芯，插件
inserting molding 嵌件成形
insoluble 不能溶解的，不能理解的
inspection 检查
inspection program 检查程序，检验程序
instance 例子，情况
instant 立即的，直接的
instantaneous 瞬间的，即可的，即时的
instruction 指令，命令
instrument 工具，器械
insufficient rigidity 强度不够
insulated runner 绝缘浇道方式
insulated runner mold 绝热流道模
insulator 绝缘体
integrated circuit 集成电路
integration 集成，综合
intellectual 有智力的，显示智力的，知识分子
intelligent node 智能节点
intend 想要，打算
intensity 强烈，强度
interactive programming 交互式编程
interactive system 交互系统
interactive 交互式的，相互作用的
interchange （指两人等）交换
interchangeable 可互换的
interchangeably 可互换地
interconnection 相互连接
interface 界面
interfere 干涉，妨碍
interference 过盈，干涉
interference fit 过盈配合，压配合
interior 内（部）的
interior cylinder 圆柱孔
intermediate 中间的，媒介，中间人，中间体
internal grinder 内圆磨床
internal grinding attachment 内圆磨头
internal stress 内应力
interpolation function 插补功能
interpolation 插补，插值法
interpolator 插补器，校对机，分类机
interpret 解释说明
inter-relationship 相互关系，内在联系
interrupted ageing treatment 分级时效处理
intersect （直线）相交，交叉
interval 间隔，空间，周期
intervention 干涉
intricate 复杂的，难以理解的
intuitive 直觉的
inventory 存货，总量
inventory 详细目录，存货，库存，总量
invert 使颠倒，使转换 n. 倒转物，倒置物
inverted blanking die 倒置落料模，倒装落料模
investigate 调查，研究
involve 包含，涉及
ionize 使离子化，电离
IPS Impact-resistant polystyrene 耐冲击聚苯乙烯
Irregular 不规则的
irregular parting line surface 不规则分型面
irregularly 不规则地
irreversible 不可逆转的，不能撤回的，不能取消的
ISO 国际标准化组织

New Words

isolating plate baffle plate；barricade 隔板
isothermal annealing 等温退火
isothermal transformation 等温转变
isotropic 等放性的
iteration 反复
itself 它本身，它自己

J

jack//n. 升降机
jaw 爪
jet molding 喷射成形
jetting//n. 喷射，注射
jiffy quick connector plug/socker 快速接头
jig 钻模
jig and fixture 夹具
JIT（just-in-time）准时生产，准时制造
jog button 手动按钮，点动按钮
jog feed rate 点动进给倍率，手动进给倍率
jog 轻推，漫步，点动 //v. 轻推，慢跑
join 连接，结合
joint 连接缝，接合处
judge 判断
jump 跳

K

kayak 皮船
key 键
keyboard 键盘
keypad 键区
kilogram 千克，公斤
kinematic 运动学（上）的
kinematic 运动学的
kinematics 运动学
kitchenware 厨房用具
knee （铣床）升降台
knock pin 顶出销
knock 打击，碰撞　n. 敲打
knockout bar 打料杆
knockout collar 打料环
knockout rod 打料杆

knockout 顶出器，脱模，卸料器
known as 称为……，把……称为
knuckle 万向接头，指节
knuckle-lever press 肘杆式压力机
knurl 节，压纹，滚花　vt. 滚花，压花

L

labor cost 劳动力成本
lack of painting 烤漆不到位
lamellar 展压，切成薄片，层状，层压，材料
laminate 层压，层合
laminated-object manufacturing 叠层物制造
laminating 切缝
lamp holder 灯架
lancing 门插销　v. 闭锁
large-quantity production 大批量生产
laser cutting machine 激光切割机
latch 锁扣
lateral 横（向）的
lateral feeding movement 横向进给运动
lathe 车床，用车床加工
latter 后面的
layout 布置图
LCP　Liquid crystal polymer 液晶聚合物
LDPE　Low-density polyethylene plastics 低密度聚乙烯塑料
lead screw 导向螺杆，丝杠
lead time 交货时间
lead to 导致
leader pin/guide pin 边钉/导边
lead-in 引入的
least 最小的，最少的，
leather 皮革
leather cloak 仿皮革
lessen 减少，减轻
letter address 字母地址
lever 杆，杠杆
lever arm 活动臂

lifter 斜顶
lifting pin 顶料销
limit switch 限位开关
line 线
line plug 喉塞
line streching，line pulling 线拉伸
line supervisor 线长
linear 直线的，线性的
linear cutting 线切割
linear motion 线性运动
linear movement 线性位移，直线运动，直线移动
linkage 连接
lip 缘，刀刃，切削刃
liquid layer curing 液层固化
liquid-crystal display 液晶显示
little finger 小指
live center 主轴顶尖
LLDPE Linear low-density polyethylene 线性低密聚乙烯
LMDPE Linear medium-density polyethylene 线性中密聚乙烯
load 负载，载荷
local heat treatment 局部热处理
located block 定位块
locating and holding device 定位和夹紧装置
locating pin 定位销，挡料销
locating plate 定位板
locating ring 定位圈
location lump，locating piece，block stop 定位块
location pin 定位销
location-limited pin 限位杆
lock pin 防转销
lock plate 锁模块
lock screw 限位螺钉，锁定螺钉
logic diagram 逻辑图
long nozzle 延长喷嘴方式

loop 环，线圈，回路，回线 vt. 使成环，以环结构
loose 宽松的，不紧的，散漫的，自由的 vt. 变松 adv. 松散地
lose contact with 与……脱离接触
loss 损失，遗失，浪费
lots of production 生产批量
low alloy steel 低合金钢
low carbon steel 低碳钢
low-cost die 简易模
lower die base 下模座
lower insulation pad 下隔热垫板
lower shoe 下模座
lower sliding plate 下滑块板
low-quantity production 小量生产
lubricant 滑润剂

M

machinability 机器加工性，切削性
Machine Ability Data Handbook 机床加工数据手册
machine building industry 机械制造业
machine centers 加工中心
machine control unit (MCU) 机床控制单元
machine tool 机床，工具机
machine tool slide 机床拖板
machinery 机器，机械
machining parameter 加工参数，切削参数
machinist 机械师，机械工人
macro 宏 adj. 宏观的，巨大的，大量使用的
made selection 模式选择
magazine attachment 机床送料装置
magazine feed 自动传输带
magazine 连式刀库，刀库，杂志，刊期
magnetic disk 磁盘
magnetic tape 磁带
magnetic 磁的，磁化的，有磁性的，磁学的，有引力的
magnetic-pulse forming 磁脉冲成形

magnetizer 加磁器
magnetostrictive 磁伸缩的
magnetostrictive oscillator 磁振荡器
main motor 主电动机
main ram packing 主柱塞组合，主压头组合
main runner 主流道
mainframe//n. 主机，大型机
mainframe-based 大型机的，主机的
maintenance 维护，保持
major defect 主要缺陷
make one' (first) appearance （首次）出现
malleability 可锻性，延展性
malleable cast iron 可锻铸铁
manganese 锰
manganese 锰
manifold 多种样式的，有许多部分的
manifold block 分流板
manifold thermocouple 热电偶
manipulate 操作，技巧地操作
manipulation 处理，操作，操纵
manipulator 操作者，操纵者，机器
manner 方法，方式
manual operation 手工操作，人工操作，手动操作
manually 手工的，手动的，人工的
manufacture 大规模制造，生产
manufacture management 制造管理
manufacture procedure 制程
manufacturing 制造业
manuscript 手稿，原稿
maraging steels 马氏体时效钢，特高强度钢，高镍合金钢
margin 刃带，边界，界限
marketing//n. 市场销售
mass 物质，质量
mass production 大量生产，批量生产
Massachusetts Institute of Technology （MIT） （美）麻省理工大学
mat 席子，垫子
mate 同事，配对，成对，成配偶
material change，stock change 材料变更
material check list 物料检查表
material removal rate 材料切除率
matrix plate 凹模固定板
matrix 矩阵，基体
matter 事件，物质
max milling cutter diameter 最大刀具直径
max tool length 最大刀具长度
maximum load on pallet 工作台最大载荷
maximum shut height 最大封闭高度
Mazak 马扎克公司
MCU（Machine Control Unit）机床控制单元，控制加工单元
MCU（Microcomputer Unit）微处理器单元
MDI（Manual Data Input）手工数据输入
meaning 意义
mechanical 机械的
mechanical drawing 机械制图
mechanical properties 力学性能，机械性能
mechanical strength 机械强度，力学强度
mechanical （toggle） lock 机械（肘节）锁紧机构
medium 介质，中间的
medium carbon steel 中碳钢
medium production rate 中等生产批量
melamine 三聚氰胺
melt temperature 熔化温度
melting stage 计量段
membrane 膜，隔膜
memory 记忆，记忆力，内存
mesh 啮合，配合
mesh with 与……啮合
mesh 网孔，网丝，网眼，陷阱
metal-effect 金属影响的，金属作用的，金属效果的
metallography 金属组织学，金相学

metallurgical 冶金学的
metalwork 金属制品，金属制造
metering 测量，计量，记录，统计结果
metric 公制的
metrologist 计量师
metrology 计量学，计量制度量衡学，度量衡，测量学
ME 制造工程师
mica 云母
microchip encapsulation 芯片封装
microcomputer 微型计算机
microconstituent 微观组织，微量成分
micrometer 测微计，千分尺
microprocessor 微处理器
microstructure 用显微镜可见的//n. 微观结构，显微结构
midfinger 中指
mildewed=moldy=mouldy 发霉
mill 铣床，铣刀，铣，铣削
miller 铣床
milling 铣削
milling machine 铣床
milling 先销
millisecond 毫秒
miniature 缩小的模型，缩图，缩影微型的，缩小的
minicomputers 小型机
minimize 最小化
minimum 最小值，最小（值）
minimum bending radius 最小弯曲半径
minor 较小的，次要的
minor defect 次要缺陷
minus 负的，减去的，负量，
misalignment 不对准，偏差
misalignment 未对准，不正，失调，不重合
miscellaneous function 辅助功能，M 功能辅助（杂项）功能
misfeed 错送，送错，传送失效

misleading 易误解的，令人误解的，误导的，另人混淆的
misplace 放错地方
missing part 漏件
mixed color 杂色
mode control 模态控制
mode transformation 模态变换
moderate 中等的，适中的
modification 更改，修改
modify 更改，修改
modifying agents 模塑剂，塑化剂
modulus 模数
modulus of elasticity 弹性模量
modulus of elasticity 弹性模量，弹性模数
modulus 模量，系数，模数
mold base 模胚（架）
mold base shut hight 模总高超出啤机规格
mold cavity 模具型腔
mold components 模具零件
mold core 模具型芯
mold flow analysis 流道电脑分析
mold for plastics 塑料成形模具（简称塑料模）
mold for thermoplastics 热塑性塑料模
mold insert 模具嵌件
mold locking force 锁模力
mold opening force 开模力
mold pressure 成形压力
mold system 模具系统
mold thickness 模具厚度
moldability 可塑性，模塑性
molding factory 成型厂
molecular [化]分子的，分子量的，由分子组成的
molybdenum 钼
moment of inertia 惯性矩
momentarily 即刻地，瞬间地
momentary contact 瞬时接触
momentary 瞬间的，刹那间的，瞬时的

New Words

monitor 班长，监听器，监视器，监控器，监控
monochromatic 单色的，单频的
monotonous feed 单向进给
monotonous 单调的，无变化的
mop 拖把
motif graphical 功能图形
motionless 不动的，静止的
mottled cast iron 麻口铸铁
mount 固定，安装
movable clamping plate 活动侧固定板，动座板
movable pilot 活动式导正销，活动式导向销
movable retainer plate 动模模板
move 运动，移动
moving core holder 活动型芯夹板，活动型芯固定板
moving platen 移动模板，活动模板
MRP（Material Requirements Planning）物料需求计划
MS-DOS（Microsoft Disk Operating System）微软磁盘操作系统
multi-component 多成分的，多组分的，多重组合的
multiple//adj. 多样的，多重的，多种多样的 倍数，若干
multiple-point tool 多刃刀具
multi-station stepped die 多工位级进模
mylar 聚脂薄膜

N

nalogue 模拟量
name of a department 部门名称
narrow 狭窄的
nature 本性，性质，自然界
near 接近，靠近，
near to 接近
nearly 将近，大约，几乎
necessitate 成为必要，需要

neodymium 钕
neodymium-yttrium-aluminium-garnet 掺钕钇铝石榴石，钕钡铅石榴石（一种固体激光
neon bulb 氖管，霓虹灯
neon 氖
nest 定位孔，窝，巢
nesting 嵌套
net weight 净重
netlike 网状的，形成网状的
neutral plane 中性面
neutral 中立的，中性的//n. 中立者，中立国，非色彩，齿轮的，空挡
nickel 镍
nickel steel 镍钢
nitriding 渗氮
No. of heater steps 电热段数
nodular graphite iron 球墨铸铁
nominal 名义上的
nonconducting 不导电的
no-reture valve 逆止阀，单向阀
normal pressure 公称力
normal pressure stroke 公称力行程
normalizing//n. & adj. 正火（的）
not up to grade n. ot qualified 不合格
not up to standard 不合规格
notation 符号，标志
notch 槽口，凹口，切痕，凹槽 v. 开缺口，冲缺口
notched wheel 棘轮
noticeable//adj. 显而易见的，值得注意的
nozzle locator 喷嘴定位套，主喷嘴
nozzle thermocouple 喷嘴热电偶
nozzle union 连嘴套
nozzle 管口，喷嘴
number of pallet 托盘数
numerical 数字的
numerical control(NC) 数字控制
numerically 用数字，数字上

185

nut 螺母，螺帽，上（拧）
nutshell 坚果壳，外壳
nylon latch lock 扣机（尼龙拉钩）

O

obey 服从，顺从
obstruct 阻隔，阻塞，遮断（道路、通道等）//n. 阻碍物，障碍物
obtain 获得
obtuse 钝的，不快的
obtuse 钝的，圆头的，愚蠢的，迟钝的
odd 奇数的，单数的
offend 犯罪，冒犯，违反，得罪，使……不愉快
oil tank capacity 油箱容量
omit 省略，疏忽，遗漏
on the other hand 另一方面
on-board 在船上（飞机，车）上，内嵌在机床上的，内置的
one stroke 一行程
one-point single-action press 单点单动压力机
on-line 在线的，即时的
OOBA 开箱检查
open-back inclinable press 开式双柱可倾压力机
opening and closing the mold 模具的开启，模具的开合
open-loop system 开环系统
open-side planer 单立柱（臂）龙门刨床
open-side press 开式压力机，单柱压力机
operation 运转，操作
operation code 操作码
operation procedure 作业流程
operation-temperature 工作温度，操作温度
operator 操作员
opposite 相对的，对面的
oppositely 在对面，相反的
optimal 最佳的，最理想的
optimize 使最优化

optimum 最佳条件，最佳状况，最佳值，最合适的，最优化的
option 选项
orbit 轨道
organic chemistry 有机化学
orientation 方向，方位，定位，倾向性，向东方
original 最初的，原始的
oscillate 波动，振动，使摆动
oscillator 振荡器
outer 外部的，外面的
outer ram 外滑块
outer slide 外滑块
outer surface 外表面
outline 外形，轮廓
overall cost 总成本
overall dimension 外形尺寸
overall 全部的，全面的，总体的
overcure 过分硫化，过度熟化，过分处治
overflow well 溢流道
overheated structure 过热组织
overlap 重叠，相交，（部分）一致，与……重叠，与……搭接
overloading 过载，超载，过负载
override 超越，占优势
overview 一般观察，总的看法
oxidation 氧化
oxide coating 氧化膜
oxyacetylene 氧乙炔的
oxyacetylene welding 氧—乙炔焊
oxynitrocarburizing 氧氮共渗

P

pack carburizing 固体渗碳
pack packing 包装
packaging tool 打包机
packaging//n. 打包
packing material 包材
pad 衬垫，衬套，垫圈，垫片
padding block 垫块

painful 费劲的，使痛苦的
painting factory 烤漆厂
painting make-up 补漆
painting peel off 脏污
pallet 棘爪，货盘
pallet change time 托盘交换时间
pallet change type 托盘交换形式
pallet shuttle 往复托盘，滑动托盘
pallet 栈板，转盘，扁平工具，棘爪，托盘
panel board 镶块
panel 面板，嵌板，仪表板
paper tape 纸带
parallel 平行的
parallel to 平行于
parameters//n. 参数
parity 奇偶校验
part from cavith（core）side 料位出上/下模
partial annealing 不完全退火
particle 微粒
particular 特别的，单个的
parting line 分型面
parting line venting 模排气
parting lock set 扣机
parting locks 尼龙拉勾
parting plane 分型面
parting surface support balance 承压平面平衡
parting 分离，断开，分形
part-program manuscript 零件编程单
pawl 棘爪
peak die load 模具最大负荷
pearlite 珠光体
pedal 踩踏板
peel 剥，削，剥落
pellet 颗粒，球粒，小球，薄皮，薄膜，表皮
pendulum-type impact test （摆锤式）冲击试验
penetration 渗透，穿透，穿透能力
pependicular 垂直的，正交的，垂线
per 按照，根据
perceive 察觉，看见
perforating 穿孔，成孔，打孔
perform 执行，完成
performance 履行，执行
perimeter 圆周，周长，周边，周边界
peripheal 外围的，边缘的，外表面的 外围设备
periphery 圆周，圆柱表面
permanent 永久的，持久的，
permanently 永久的，永存的，不变的
permit 许可，允许
perpendicularity 直度
perpheral device 外围设备，外围装置
personnel resource department 人力资源部
personnel 人员，职员，人事部门
perspective 透视，透视法，透视画
pertinent 有关的，相干的，中肯的
phase change 相变
phase transformation diagrams 铁碳合金相图
phenol formaldehyde 苯酚甲醛
phenol formaldehyde resin 聚酯树脂
photoplymerization 光化聚合作用，光化聚合
photopolymer 光敏聚合物
physical properties of a material
physical property 物理性能
physical-model-producing technology 实体造型生产技术
pick up 拾起，采集
piercing die 冲孔模
pigment 色素，颜料
pillar 柱子，柱状物 vt. 用柱支撑，用柱加固
pimple 丘疹，面泡，疙瘩

pin gate 针点浇口
pin 镶针 针，钉，栓，销
pinch-type clamp 夹起式夹钳
pinion 小齿轮
pin-point gate 细水口
pipe plug 喉塞
piston type 活塞型的
piston 活塞
pitch diameter 节圆直径
pixel （显示器或电视机图像的）像素
place 地点，位置，放置
planer 龙门刨床
planner 计划者
planning 刨削
planning department 企划部
plant 植物，庄稼，工厂，车间，设备，布置
plasma panel display 等离子面板显示
plasma 等离子体，等离子区
plastic 塑料的，可塑的
plastic basket 胶筐
plastic deformation 塑料变形
plastic parts 塑胶件
plastic tube 塑胶管
plasticity 可塑性，塑性
plasticization 增塑，塑化，塑炼
plasticize 使成可塑体，使……塑化
plasticizing unit 塑化装置
plate 金属板，压条
platen 压盘，台板，隔板，压板
pliable 可塑的，可锻的，柔韧的，柔顺的
plotter 绘图机
plunge-cut grinding 横向全面进给磨削
plunger bush/plunger bushing 浇口套
plunger injection system 柱塞式注射系统
plunger 拄塞，潜水者，活塞，短路器
plus 正的，略大的，正量
ply bar scot 开模槽
pneumatic screw driver 气动起子

pneumatic 气动的，有气胎的，汽力的，风力的，压缩空气推动的
pocket 口袋，袋//adj. 袖珍的，小型的
poeder 粉，粉末，火药，药粉，尘土 磨成粉，使成粉末
point 点，刀尖，钻尖
point-to-point 点对点
point-to-point tool movements 刀具的点位运动
polishing/surface processing 表面处理
polyacrylate 聚丙烯酸酯
polycarbonate [化]聚碳酸脂
polyester resin 酚醛树脂
polymer 聚合体，聚合物
polymeric resin 聚合树脂
polymerization 聚合
polymerize （使）聚合，聚合，重合
poor incoming part 事件不良
poor processing 制程不良
poor staking 铆合不良
porosity 多孔性，有孔性，空隙度，多孔结构
portable transfer mold 移动式压注模
portion 部分
position 职务
positioning system 定位系统，布置系统
positioning with scales 配光栅尺定位精度
positioning/full stroke 定位精度、全行程
positive 正的，肯定的，阳性的
positive mold 不溢出式压缩模
positive mold 挤压式模具
positive plunger type 正柱塞型
positive plunger-type transfer mold 正柱塞型传递模
positively 确定，确实地，正
positive-motion drive 强制转动
possibility 可能性
post screw insert 螺纹套筒埋值
post 柱，杆，桩 vt. 告示，揭示

New Words

post-processing 后处理（的），后加工（的）
potential 潜能，潜力
pounds per square inch 磅/平方英寸
pouring temperature（rate，time）浇注温度（速度，时间）
power button 电源按键
power wire 电源线
powerful 强有力
PPS（Production Planning and Scheduling）生产计划表，产品计划表
practitioner 从业者，开业者
precious 宝贵的，贵重的，珍爱的
precise 精密的，精确的
precision 精度
predetermine 预定，预先确定
pre-dry 预干燥（的）
prehardened steel 预硬化钢，预淬火钢
preliminary 预备的，初步的
preliminary drawing 一次图，初步图，草图
preliminary（final）mold design 初步（正式）模图设计
premature failure 过早破坏，过早失效
premeasure 预测量，预测度，预调节
preparatory function 准备功能
prepare for，make preparations for 准备
preprocessing 前处理（的），生产前（的），试制（的）
prerequisite 先决条件 adj. 首要必备的
present 介绍，呈现，提出，
preshape 预成形
president 董事长
press 压，冲压，压力机
press bed 压力机床身
press brake 弯扳机，剪扳机
press tonnage 压力机吨位
press 压，按，压力，压力机，新闻压，逼迫，受压
pressure die-casting dies 压力铸造模具
pressure of air cushion 气垫压力
pressure plate=plate pinch 压板
pressworking 压力加工
pressworking 压力加工，冲压加工
previously 先前，以前
price/performance 性价比
primary 第一位的，初级的
primary motion 主运动
principally 主要地
principle 原理，原则，主要的，
prismatic 棱镜的
procedure 规格，规程，程序
proceed 进行，发生
process parameter 成形工艺参数
processing parameter 加工参数
processing，to process 加工
product development cycle 产品研发周期，产品开发周期
product quality 产品质量
production capacity 生产力
production department 生产部门
production line 流水线
production rate 生产率
production run 生产（运行）周期
production schedule 生产计划表，生产进度表
production unit 生产单位
production，to produce 生产
productivity 生产率，生产能力
professional 专业人员 adj.专业的，职业的
profile 剖面，侧面，外形，轮廓
programmer 编程员
progressive die 级进模，连续模
progressive die，follow（-on）die 连续模
projected area 投影面积，投影区域
prolong 延长，拖延
prominent 突出的，卓越的，显著的，凸出的，突起的，杰出的

189

pronounced 讲出来的，显著的，断然的，明确的
propagate 繁殖，传播，宣传
propeller 推进者，推进器
proportion 比例，部分
proportional 比例的，成比例的，相称的，均衡的
proposal improvement 提案改善
prototype part/workpiece 样件
prototype 原形，模型，样品，样机，范例，均衡的
provision 供应，（一批）供应品，预备，防备，规定
pull-down menus 下拉式菜单
pulley 皮带轮，滑轮
pulsate 脉动，波动
pump motor 液压泵电动机
punch 冲床，冲孔，穿孔
punch bushing 凸模护套
punch holder 冲头夹紧器
punch press，dieing out press 冲床
punch 冲头，凸模，冲压机，冲床，打孔机 vt. 冲孔，打孔
punched hole 冲孔
punched tape 穿孔带
punch-holder plate 凸模固定板，凸模夹持器
punching 冲孔，穿孔
punching machine 冲床
push bar 推杆

Q
QC 品管
QC Section 品管科
qualified products，up-to-grade products 良品
quality assurance system 质量保证体系
quality control department 质量控制部门，质量管理部门
quality 品质

quenching 淬火
quick-change tool 快换刀具
quill 衬套，套管轴，羽毛 vt. 卷在线轴上

R
rachet（防倒转）的棘齿
rack，shelf，stack 料架
radial 半径的
radial drilling machine 摇臂钻床
radial 径向的，放射状的，半径的
radially 沿径向
radius 半径
rag 抹布
rail 轨，铁轨
rake 前角
ram 滑枕
RAM（random-access memory） 随机存取器，随机存储器
ram 冲头，压头，柱塞，撞杆，滑枕，滑块，随机存储器
random 任意的，随便的
randomly 任意地，随便地，随机地
ranges fom 在……范围，从……到范围之内
rapid 快速的，快的
rapid tooling 快速制模，快速准备工装夹具
rapid travel override 快速进给倍率开关
rapid traverse 快速进给
rapidly 快，迅速的
rapid-prototyping machinery 快速成形机
ratchet 棘齿
rate 率，比率，速度
rather than 而不
ratio 比，比率
raw material 原材料
reach 到达，达到
reaction 反应，反作用
reaction-injection molding 反应注射成形

reactive molding 反应成形
reactive 活性的，反应的，起反作用的，反动的，电抗性的
reak-time 实时的，即时的
real-time fault recovery 实时纠错恢复，实时故障恢复
real-time FMS control 实时柔性制造系统控制
ream 铰孔
reamer 铰刀
reaming 铰刀
rear plate 后板
rear platen 后板，背板
receive 领取
reciprocate 来回往复运动
reciprocate 使往复运动，互换，互给
reciprocating-screw system 往复螺旋式系统
reclaim 回收，再生，重新使用，改良，改造，革新，复原
recognize 承认，认识
recommend 推荐，介绍
recommendation 推荐，劝告，建议
recrystallization temperature 再结晶温度
rectangular worktable 矩形工作台
rectangular 长方形的，矩形的，成直角的
recycle 使再循环，反复应用//n. 再循环，再生，重复利用
reducing die 缩口模，缩径模
reel-stretch punch 卷圆压平冲子
refer to 指的是，称为
reference 参考，基准面
reference coordinate system 参考坐标系
reference datum（plane）参考平面
reference reture 基准复位，零点返回
reference surface 基准面，参考面
regard 认为，考虑
regard…as 认为……是
regardless of 不管，不顾

registration 登记
registration card 登记卡
regrinding 重磨
regular 规则的，对称的
regulation 整顿
reheating 重新加热
rein-forced 增强的，加强的
reinforced-plastic 增强型塑料的，加强型塑料的
relationship 关系
relative to(be) 相对于，关于
release 释放，解放，放松，放出 释放装置，排气装置
relevant 有关的
relive 减轻，解除，援救，救济，换班
remedial//adj. 治疗的，补救的
remedy 药物，治疗法，补救，赔偿 治疗，补救，矫正，修缮，修补
removal 移动，切除
remove rifle 旋转，运行
remove 移动，消除，去除
render 呈递，归还，着色，交纳，粉刷
repair 修理
repeatability 可重复性，反复性，再现性
repetitive 重复的，反复性的
replace 替换，代替，把……放
report 报告，通报
representative 典型的，有代表性的
reservori 水库，蓄水池，容器，储存器，储藏
reset button 重置键
reset 重新安排，重新设置，复位
residence time 驻留时间
resilience 有弹力，恢复力
resin transfer molding 树脂传递成形
resin 树脂，松香，树脂制品，涂树脂于……
resizing 尺寸再生
resort 求助，诉诸，采取手段，凭借，手段

respective 分别的，各自的
respectively 分别地，各自地
restrict 限制，约束，限定
restriking die 矫正模，校正模，整修模
result 结果
result from 起因，由来
result in 导致，结果形成
resultant 作为结果而发生的，合成的
retrieval 取回，恢复，修补，重获
retrieve 重新得到，重新找到
return 返回，退回
return pin 复位杆
return pin and cavity interference 回针碰料位
return stroke 返回行程，回程
return-blank type blanking die 顶出式落料模
reverse 相反，背面，倒退，
reverse angle = chamfer 倒角
reverse core 反核心
reverse draw 逆向拉深，反向拉深
reverse 相反，背面，反面，倒退 相反的，倒转的，颠倒的，颠倒，倒转
reversely 相反
reversible 可逆的
review 重查，评论，回顾
review 回顾，复习，浏览，回顾，复习，评论，浏览
revision 修正，修改
revision 修订，修改，修正，版本
revolve （使）旋转
rheological 流变学的
rib 肋，肋骨
ribbon punch 压筋冲子
rifle 步枪
right-hand rule 右手定则，右手法则
rigid 刚性的
rigid pilot 固定式导正销，固定式导向销
rigidity 刚性，硬度

rind//n. 外壳，外表，硬层，果皮
ring 环，圈
ring finger 无名指
ring gate 环形浇口
rivet 铆钉，铆接，铆
rivet gun 拉钉枪
rivet table 拉钉机
robot 机械手
robotics 机器人技术
robustness 精力充沛的
rocker mechanism 摇杆机构
rocker-arm press 摇臂式压力机
Rockwell test 洛氏硬度试验
rod 杆，棒，连杆，拉杆，推杆，测杆
roll 轧制，滚轧，卷轴，
roll material 卷料
roller 滚筒，辊子，导轮（用在线切割机床上）
rolling 轧制，滚轧
ROM（read-only memory） 只读存储器
rotate 回转，转动
rotating speed, revolution 转速
rotating-screw feeder 螺旋式螺纹紧定器，螺旋式螺纹起子
rotation 回转，转动
rotational 旋转的，循环的
rotational molding 旋转成形
rough turn 粗车
round 圆的，围着，围绕着
round punch 圆冲子
round（full/half runner）圆形流道
row 排，列
RP and M（Rapid Prototyping and Manufacturing）快速成型制造
rpm=revolution per minute 转/分
rubber 橡胶
rubber forming 橡胶成形
rubber-loaded strippers （硬）橡胶加载卸料板

New Words

ruby 红宝石
runner 横浇道
runner balance 流道平衡
runner bar 浇口棒
runner less 无浇道
runner less mold 无流道模
runner plat 浇道模块
runner plate 流道板
runner stripper plate 浇道脱料板
runner system 浇道系统
runner 流道
rust 生锈

S

S.H.S.B 栓打螺丝
saddle 鞍，鞍状物，滑板
safe stock 安全库存
safety rachet bar 安全（防倒转的）棘齿拉杆
safety 安全
satisfactory 满意的，符合要求的
scale 氧化皮，测量，起氧化皮
scanner 扫描器，扫描仪
scar 结疤，疤痕 v. 结疤，使留下伤痕，创伤
schedule 进度表
schematic diagram 示意图
schematic 示意性的，图解的，按照图解的，纲要的
Science Research Council 科学研究协会
scope （活动）范围，机会，余地
scrap 废品，废料
scrap rate 废品率
scrap 小片，废料，残余物，废料碎片，扔弃，敲碎，零碎的，废弃的
scrape 刮，削
scrapless progressive die 无肥料连续模
scratch 划痕，刮伤，划伤
screw diameter 螺旋杆直径
screw driver 起子

screw driver holder 起子插座
screw L/D ratio 螺杆长度比
screw press 螺旋式压力机
screw rotation 螺杆转速
screw 螺丝
seal 封，密封
sealed ring 密封圈，密封环
sealed 未知的，密封的
sealing 密封
seamless integration 无缝集成
seamless//adj. 无缝合线，无伤痕的
section supervisor 课长
secure 保证，获得，可靠的，安全的，可靠的，放心的，无虑的，保护，保证，使安全
segment 段，节，片段，分割
select 选择
selective laser sintering 选择性激光烧结
selector 选择器
selector lever 选速手柄
self 自己，自行
self tapping screw 自攻螺丝
self-aligning ball bearing 自位滚珠轴承
self-contained 独立的，配套的
self-hardening steel 自硬钢
semi-permanent 半永久的
semipositive mold 半溢出式压缩模
sensitive 敏感的
sensor 传感器
separate 分离的，个别的，分开
separation 分离，分开，分类，间隔
separator 隔离罩，隔板，分离器
sequence 继（连）续，一连串
sequence number 顺序号
servo control system 伺服控制系统
servo 伺服，伺服系统
servomenchanism 伺服机构，伺服系统，自动控制装置
servomotor 伺服电动机，伺服马达

set up time 安装时间
setback 顿挫，挫折，退步，回退
shaft 轴
shallow 浅的，浅薄的
shank 钻柄，模柄
shank-hole size 模柄孔尺寸
shankless die 无柄模具
shaper 牛头刨床
sharpen 使锐利，变尖
sharpness 锐利，尖锐
shaving die 切边模，修边模
shear 剪（切）
shearing force（plane）剪切力
shearing force diagram 剪力图
shearing 剪切，剪除
sheath heater 发热管
sheath 鞘，护套，外壳
shedder 卸件装置，推料机
sheet 张，片，薄钢板
sheet metal 板料
sheet metal forming 钣金，金属板成形
sheet metal parts 冲件
shiver 饰纹
shock 冲击，振动
shock resistance tool steel 抗冲击工具钢
shop floor 车间
shortcoming 缺点，短处
shot chamber 注射室
shot volume 注射量，压注量
shot 注射，注入，范围，射程
shoulder guide bushing 中托司
shoulder 肩，肩部，侧翼 vt. 肩负，承当
shrinkage 收缩，收缩量，收缩率
shrinking/shrinkage 缩水
shut die 架模
shut height 闭合高度
shut height of a die 架模高度
shut height of press machine 压力机闭合高度

shut&n. 关上，闭上，关闭，合拢
shuttle table 移动工作台
shuttle 航天飞机，梭子，穿梭，往复运动，穿梭往返，使穿梭般来回移动
side clearance angle 侧隙角
side core 侧型芯
side edge 侧刃
side gate 侧浇口
side holes 侧孔
side locating face 侧定位面
side-push plate 侧压板
Siemens 西门子公司
significant 重大的
significant 有意义的，重大的，重要的，明显的，显著的
silicon chip 硅片
similar to 类似于……的
simple 简单的
simplify 简化，单一化，简单化
simulation 仿真，假装，模拟
simultaneous 同时的，同时发生的
simultaneously 同时地，同步地，同时发生地
single block 单程序段
single blow 一击，一下子
single execution 单步执行
single operation 单步运行
single point tool 单刃车刀
single-acting main ram 单向主压头，单向主柱塞，单向主撞杆
single-action straight-slide eccentric mechanical press 闭式双点偏心轴单动机械压力机
single-point cutting tool 单刃刀具
single-stage plunger 单级柱塞
single-stage plunger machine 单级柱塞注射机
single-station piercing die 单工序冲孔模，单工位冲孔模
sink mark 缩影，缩形

New Words

sink 沉下，（使）下沉
sintered carbide 硬质合金
sintering 烧结
size 大小，尺寸，加工到
slag well 冷料井
slant （使）倾斜，歪向 n. 倾斜
sleave 套筒
sleeve 袖套，套管，给……装套筒
slice 薄片，切片，一份，部分，片段
slide （使）滑动，（使）滑行，
slide bedway 滑动床身导轨
slide motor 滑块电动机
slide surface size 滑块底面
slide 行位（滑块）vt. 使滑入 n. 滑动，滑块滑座，冲头
slider 滑块，滑雪者，滑冰者
sliding block 滑块
sliding dowel block 滑块固定块
sliding rack 滑料架
slippage 滑动量，滑程
slipped screwhead/shippery screw thread 滑手
slipped screwhead/slippery screw head 螺丝滑头
slippery 光滑的
slit 切开，切口，切缝
slit gate 缝隙浇口
slot 缝，狭槽
sludge 软泥，淤泥，矿泥，煤泥
slug 弹丸，金属小块，嵌片，冷料
slurry 泥浆，浆
smoothly 顺利
snap ring 卡环
snap 卡扣，夹子，快门
so-called 所谓的，号称的
socket head cap bolt 沉头螺钉
socket 沉头，穴，孔，插座，牙槽 v.给……配插座
sodium 钠

soften （使）变柔软，（使）变柔和，使……软化
software package 软件包
software 软件
solder 焊料
soldering 锡焊，软钎焊
solid model 实体造型，实体建模
solid molding 实体建模
solid-based curing 实体基固化
solidification 凝固
solidify （使）凝固，巩固
somewhat 稍微，有点
sophisticate//vt. 使复杂，曲解，使精致，掺和，弄复杂
sophisticated 复杂的
sophisticated 复杂的，高级的，成熟的，精致的，老练的，诡辩的，久经世故的
SOP 制造作业程序
sound 健全的，坚固的
space between bars 柱间间隔
space-age 空间时代
spacer 衬套，衬垫，垫片
spacer block 间隔块，支撑块
spacer plate 分流板腔，间隔板
spak 火花，火星，闪光，电火花，瞬间放电
spare dies 模具备用品
spare parts=buffer 备件
spark erosion machining 电火花加工法
spatial 空间的
spear head 刨尖头
special assistant manager 特助
special shape punch 异形冲子
specialize 专攻，专门研究，使适应特殊目的，使专用于
specific 特殊的
specification 技术要求，详述，规格，说明书，规范
specimen 样品，试样

speckle 斑点
spectrum 光，光谱，型谱，频谱
speed discrepancy 转速差
speed ratio 速比
spelter 硬钎焊料，锌铜焊料
spelter solder 硬焊料
sphericity 成球形，球（形）度
spheroidized structure 球化组织
spheroidizing 球化处理
spindle 轴，主轴，连接轴
spindle bore diameter/draw tube diameter 主轴通孔/拉管内径
spindle center line to pallet surface 主轴中心至工作台面距离
spindle drive motor 主轴电动机功率
spindle max. torque 主轴最大扭矩
spindle nose 主轴鼻端形式
spindle nose to table center line 主轴端面至工作台中心距离
spindle speed 主轴转速
spindle taper 主轴锥孔
spindle 阀针主轴，轴，转轴
spinning 旋压
spin-off 派生的
spiral 螺旋形的，螺纹的
spiral groove 螺旋槽
split mold 分割式模具
spot check 点检，抽查
spot facing 锪端面
spot welding 点焊
spreader 分流锥
spring 弹簧，弹回
spring box 弹簧箱
spring compressed length 弹弓压缩量
spring rod 弹弓柱
spring 弹簧
spring-back 回弹，弹复
spring-back angle 回弹角
spring-box eject-plate 弹簧箱顶板

spring-box eject-rod 弹簧箱顶杆
spring-loaded strippers 弹簧载入卸料板
sprue 直浇道，主流道，浇口
sprue base（bush, gate, puller）直浇道窝（浇口套，直接浇口，拉料杆）
sprue bush/sprue bushing 浇口套
sprue bushing 唧嘴
sprue diameter 唧嘴口径
sprue gate 射料浇口，直浇口
sprue gating 中心浇口，浇道浇口
sprue less 无射料管方式
sprue lock pin 料头钩销（拉料杆）
sprue puller/sprue puller pin 拉料杆
sprue 浇口，主浇口，主入口，流道，主流道，竖流道，直流道
spur 刺激，齿
spur gear 直齿轮
spure sperader 分流锥
squareness 方（形），垂直度
stacking 堆积，堆栈，堆叠
stage die 工程模
stainless 不锈钢
stainless steel 不锈钢
staker=reviting machine 铆合机
stamp 冲压
stamped punch 字模冲子
stamping 冲压
stamping and punching die 冲模
stamping factory 冲压厂
stamping-missing 漏冲
stand 站，经受
stand up to 经受（得住），耐
standard 标准
standard keyboard 标准键盘
standard parts 标准件
standardize 标准化
standpoint 立场，观点
starting stop 始用挡料销
stationary platen 固定模板，静止模板

New Words

statistics 统计
steadiness 稳定性，坚定，稳固，平衡，稳态
steak 条纹，条痕，纹斑，色线
steam 蒸汽
steam engine 蒸汽机
steam hammer 蒸汽锤
stearine 甘油，硬脂
steel belt 钢带
steel plate 钢板
stepper bolts 梯形螺钉
stepper motor 步进电动机
stepping bar 垫条
stepping motor 步进电动机
stereolithography 立体平板印刷术
sticking 扣模
stiff 刚（性）的
stiffening rib punch = stinger 加强筋冲子
stiffness 刚度，强度
stimulate 刺激，激励
stock stop 挡料销，挡料块
stock 树干，杠杆，托柄，原料，材料台，座，给……装上把手
stop collar 止动垫圈，止动环
stop pin 垃圾钉
stop plate 挡板
stop/switch off a press 关机
stopper 阻挡器
straight 直的，直线的
straight cut 直线切割
straight guide bushing 直导套
straightforward 坦率的，简单的，易懂的，坦率地
straight-line 直线
straightness 直线度
straight-side single-action double-crank press 闭式双点单动双曲柄压力机
strain 张力，应变　v. 拉紧，扯紧，（使）紧张

strategy 策略，战略
streak 条纹，条痕，纹斑，色线
streamline 流线型的
strength 强度
strength 力（量），强度
stress crack 应力电裂
stress 应力，压力，紧迫，重点，强调，重音，强调，重读，使受应力
stretch 伸展，伸长
stretching 伸，拉伸，伸长，延展，展宽
strip 条钢，带钢，狭条
条，带，条料，带料
stripper force 顶料力，卸料力
stripper plate 推板，卸料板
stripper 卸料器，卸料板，剥离器
stroke 冲程，行程
stroke length 滑块行程
stroke of cushion 气垫行程
strole 打，击，敲，冲程，行程，划短线于，删掉
stud 销，螺栓，螺柱
subarine gate 潜入式浇口
subassembly 组合件，部件
subdivide 再分，细分
subject 题目，主题，科目，从
subject to 受……支配的，以……
sub-line 支线
submarine gate，tunnel gate 潜伏浇口
subsequent 后来的，并发的
subsequently 其后，接着
substance 物质
substantially 充分地，实质地，真实地
substitute 代替者，代替，替换
subtraction 减少
subtractive processes 去除材料过程
subtravtive 减去的，减出的，去除的
subzero 深冷处理
successive 继承的，连续的
suck 吸，吮，吸取，吸附

197

sufficient 充分的，足够的
sulfur 硫磺
summarize 概述，总结，摘要而言，摘要
summation 总结，总和，总数，加法，求和
superimpose 添加，双重
superimposition 叠印
superplastic forming 超塑成形
supervision 监督，管理
supervisor 督导人
supplier 使应者，补充者，厂商，供给者
supply 提供，电源，提供，
supply…with(for, to) … 把……供给……
support capacity for chuck/center 最大夹持重量（卡盘/卡盘+顶尖）
support pillar 撑头
support plate 托板
suppress 镇压，抑制，查禁，使止住
surface coating 表面涂覆，表面涂层
surface finish 表面粗糙度
surface grinder 平面磨床
surface molding 曲面建模
surplus 过剩，剩余
surplus stock 余量
suspend 吊，悬挂
延迟，悬挂，延缓
suspension 暂停，中止，悬而未决，延迟
sustain 支持，维持
swaging 型锻，模锻
swarf 金属屑；塑料屑
sweeper 扫把
swing over bed 台面旋径
swing over carriage 滑鞍上最大回转径
switching runner/gate 转水口
swivel 旋转
swollen 肿胀的，鼓起的，膨胀的
symbol 记号，符号
symmetrical 对称的，均匀的

synchronize 同步
synthetic fiber 合成纤维
synthetic resins 人造树脂，合成树脂
synthetic 合成的，人造的，综合的合成剂
system 系统，体系，制度
system prompt 系统提示
system variable 系统变量

T
tab gate 搭接浇口
table 工作台
table feed 工作台进给
table increment 工作台分度
table index 360 position 工作台360位置回转定位精度
table index speed/90deg 工作台90度分度速度
table load 工作台承载
table longitudinal movement 工作台纵向移动
table working surface 工作台尺寸
tablet 标牌，小平板，药片，压片
tableware 餐具
tailored 剪裁讲究的
tailstock spindle diameter 套筒直径
tailstock spindle stroke 套筒行程
tailstock 尾架，尾座，顶针座
take care of 照看，留心，处理
take…into consideration 对……加以考虑，
taker 取料机
taking into account 把……考虑在内，考虑到，注意到
tang 柄舌
tap 攻丝螺纹
tape format 纸带格式
taper 锥形，锥度，逐渐变细，锥拔
tapping 攻丝，攻螺纹，穿孔
technical manual 技术手册，技术指南
tedious 单调乏味的
tempering 回火

New Words

temporay 暂时的，临时的，临时性
tensile and compressive stresses 拉应力和压应力
tension 张（拉）力，拉伸
ten-station turret center 十工位车削加工中心
term 术语
terminal//n. 终点站，终端，接线端//adj. 终端的，每学期的
termination 终止
terminology 术语，专门名词
terms 术语
testing and inspection equipment 测试和检测设备
texturing 组织的，纹理的，网纹的
the Electronic Industries Association (EIA)
the former 前者
the latter 后者
themosetting 热固性的，热硬性化的热塑性塑料
theoretic shot volume 理论射出量
theoretical 理论的
thermal contraction/shrinkage 热收缩
thermal properties 热性能
thermal 热（量）的，由热造成的//n. 上升暖气流
thermocouple 热电偶
thermocouples 探针
thermoforming 加热成型
thermoplastic mold 热塑性模具
thermoplastic 热塑性的 n. 热塑性塑料
thermosetting mold 热固性模具
thermosetting resin binder 热固树脂黏结剂
thero expansion 热膨胀
thickness gauge 厚薄规
thickness of air bolster 垫板厚度
thoroughly 完全地，彻底地
thread 车螺纹
threading 攻丝，车螺纹
threading cycle 加工螺纹循环
three plate 三极式模具
three-dimensional surface 三维曲面
three-jaw(universal)chuck 三爪（万能）卡盘
three-plate pin-point gate injection mold 三块板式点浇口注射模具
throat depth 喉深
thrust 推力，牵引力
thrust type of bearing 推力轴承
thumb 大拇指
tie bar/rod 连接杆，拉杆，横梁
time-consuming 费时的
times of stroke per minute 行程次数
time-saving 省时的
tip 顶，尖端，嘴尖，末端
tire 疲劳，劳累，轮胎 vt. 使疲倦，使厌烦//vi. 疲劳，厌倦
to apply oil 擦油
to bending 折弯
to clean a table 擦桌子
to clean the floor 扫地
to collect，to gather 收集
to compress，compressing 压缩
to connect material 接料
to continue 连动
to control 管制
to cut edges=side cut=side scrap 切边
to draw holes 抽孔
to feed，feeding 送料
to file burr 锉毛刺
to fill in 填写
to fix a die 装模
to grip（material）吸料
to impose lines 压线
to load a die 装上模具
to load material 上料
to looser a bolt 拧松螺栓
to lubricate 润滑

to mop the floor 拖地
to move away a die plate 移走模板
to move，to carry，to handle 搬运
to notice 通知
to pull，to stretch 拉伸
to put forward and hand in 提报
to put material in place，to cut material，to input 落料
to repair a die 修模
to return material/stock to 退料
to return of goods 退货
to reverse material 翻料
to revise，modify 修订
to send delinery back
to stake 铆合
to stake，staking，reviting 铆合
to start a press 开机
to stock，storage，in stock 库存
to switch over to，switch…to，throw…over，switching…over 切换
to take apart a die 卸下模具
to tight a bolt 拧紧螺栓
to unload material 卸料
toggle 肘节，肘环套接，曲柄杠杆机构，曲拐，触发器
tolerance of fit 配合公差
tolorance 公差
tonnage of press 压力机吨位
tonnage 吨数，吨位
too…to 太……以至于不能
tool bit 刀头
tool change time 换刀时间
tool changer 换刀
tool head 刀具头架
tool life 刀具寿命，刀具耐用度
tool magazine 刀具库
tool magazine capacity 刀库容量
tool offset 刀具偏置
tool post 刀座，刀架
tool shank 刀柄形式
tool steel 工具钢
tool-holder 刀夹，刀柄，刀杆，刀具夹持器
tooling 工夹具，工艺装备
tooling charge 工具（刀具）费用
tooling cost 加工成本
top clamp plate 上模侧固定板，定座板
top stop 上死点
top-drive sheet-metal stamping automatic press 上传动板料冲压自动压力机
toughness 韧性，韧度
toughness 韧性，韧度，塑性
toy 玩具
trailer=long vehicle 拖板车
transducer 传感器，变频器，变换器，转换器
transfer 移动，传递，转移
transfer machine 传送机
transfer mold 传递模，压铸成型模
transfer molding 传递成型，压铸成型
transfomer 变压器
transition 转变，转换，跃迁，过渡，变调
transmission gearing 齿轮转动机构
transmission rack 输送架
transmission 播送，发射，传动，传送，传输，转播
transparent 透明的，显然的，明晰的
transportation 运输
transverse 横向的，横梁
trap 圈套，陷阱，闸门 vt. 收集，止住
travel dimension 行程和加工范围
traveling-column 行程立柱
traveling-wire EDM machine 线电极电火花加工机床，慢走丝线切割加工机床
traverse adjustment 横向调节
traverse control 横向控制
traverse 横过，通过，经过，横贯，横断，横断的，来回移动的

New Words

traverse-cut grinding 横向磨削
tremendous 极大的，惊人的
trial run 试运行，试生产
trigonometry 三角法，三角学
trillion 万亿
trim die 翻边模
trimmed 平衡的
trimmed-off 清理，去掉
trimming press 切边压力机
trimming punch 切边冲子
trimming 清理，除去
trip 往返，行程，脱开，切断，解扣，断路，有活力
triple 三倍的，三重的，三联的 使增加三倍
triple-action press 三动式压力机
trolley 台车
trouble-shooting 发现并修理故障（的），解决问题（的）
tube 喉管
tube heater 发热圈
tungsten 钨
tunnel gate 隧道式浇口
turn 车削
turning center 车削中心
turret press 冲模回转压力机
turret 转台，转塔，转塔刀架
twin-drive press 双边齿轮驱动压力机
twist 扭转
twist drill 麻花钻
two plate 两极式（模具）
two-plate mold 两块板模具
two-point mold 双点压力机
two-point single-action press 双点单动压力机
typical 典型的，标准的

U
ultimate strength 极限强度
ultimate tensile strength 极限抗拉强度
ultimate 最后的，最终的，根本的，最终
ultimately 最终，最后，终于，根本，基本上
ultrasonic machining installation 超声波加工装置
unanticipated 不曾预料到的，未曾想到的
unattended 没人照顾的，未被注意的
uncoiler & straightener 整平机
undercut 凹槽，潜挖，底切，砍口 挖凹槽，底切，雕出，潜挖
undergo 经历，遭受，忍受
underlying 在下面的，根本的，潜在的
undue 过度的
unfold 打开，显露，开展，阐明，伸展
unidirectional 单向的，单向性的
uniform 统一的，相同的，一致的，始终如一的，均衡的
uniformity 均匀性，一致性
unify 统一，使成一体
unilateral 单向的
unique 唯一的，独特的
unique to ……所独有的
uniueness 唯一性，单值性，独特性
universal 万能的，通用的
universal grinding machine
unmanned operation 无人化操作
up milling 逆铣
update 使现代化，修正，校正，更新 现代化，更新
upper die base 上模座
upper holder block 上压块
upper insulation pad 上隔热垫板，上隔热垫片
upper mid plate 上中间板，定位支承块
upper padding plate blank 上垫板
upper shoe 上模座
upper supporting blank 上承板
upstroke 上行程，向上行程
use beryllium copper insert 用铍铜做镶件

user interface 用户界面
utmost 极度的，最远的

V

vacancy 空缺
vacuum cleaner 吸尘器
valve gate 阀门浇口
valve pin 伐针
valves 气阀
vanadium 钒，铅矿
vaporize （使）蒸发
variation 变化，改变
various 不同的，多样的
vaseline 凡士林
vee V 字形，字形的
vendor 卖主
ventilation 排气，通风
verification 验证，确认，查证，作证
versatile 多用途的，多方面的
versatile press 通用压力机
versatile 通用的，万能的，多才多艺的，多面手的
versatility 多功能性
version 译文，译本，翻译
versus 与……相对
vertical 垂直的
vertical machining center 立式加工中心
vibration 振动，摇动
vice versa 反之亦然
vickers test 维氏硬度试验
violation 违反，违背，妨碍，侵害
virtual prototyping 虚拟原型成型
virtual 虚的，虚拟的，实质的
virtual-reality 虚拟现实的，虚实的
virtue 优点，功效
viscous 黏性的，粘滞的，胶粘的黏质，黏性
vise 虎钳，台钳，夹具
visualization 可视化
visualizer 视觉型的人，观察仪

vitally 非常
volatile 挥发性的，可变的，不稳定的，爆炸性的//n. 挥发物
voltage 电压
voltage switch of SPS 电源电压接拉键
volumetric 测定体积的，体积的，容积的，容量的
VS=versus 对，与……相对

W

warm runner mold 温流道模
warm runner plate 温流道板
warpage 翘曲，扭曲，热变形
washer 垫圈，洗衣机，洗碗机
washing-marching agitators 洗衣机搅拌器
waste 废料
watch press 台式压力机
water line 运水
water line interferes with ejector pin 顶针碰运水
water spots 水渍
wax 蜡，蜡状物，蜡制的
wear 磨损，损坏
wear pads 耐磨垫圈
wear resistance 耐磨性
wedge 楔，斜铁
wedge wear plate 耐磨板/油板
wedge 楔块，楔形物，劈，浇口，楔入，楔进
WEDM（Wire Electrical Discharge Machine）线切割加工机床
weld 焊接，熔焊 welding 焊接
welder 电焊机
welding 焊接
welding line 熔合痕
welding mark 熔合痕
well 好
well type 蓄料井
wet station 沾湿台
whether…or 无论……或……

New Words

whilst=while 时时，同时
winding 绕，缠，绕组，线圈
wire drawing 拉丝
wiring diagram 布线图，线路图
with respect to 关于，至于，就……而论
with the help of 借助于，在……帮助下
withdraw 收回，撤销//vi. 缩回，退出
withstand 抵抗，经得起
wobble 摇晃，摇摆
word address 字地址
work cell 工作间
work out 解决，算出，加工出
working drawing 工作图
working gap 加工间隙
working medium 工作介质，加工介质

workshop 车间，工场
worktable 工作桌
worm 蜗杆
worm gear 蜗轮
wrinkling 起皱 a great deal of 大量的
wrinkling 起皱现象，起皱纹
wrist//n. 手腕，腕关节 X
wrong part 错件

Y

yoke 刀杆支架
zinc chloride 氯化锌
zinc chloride 氯化锌 yttrium 钇
zoom 窗口，图像电子放大
放大，使摄像机移动

References

[1] 杨成．模具专业职场英语．北京：电子工业出版社，2009．

[2] 夏虹，等．专业英语（机械类用）．北京：机械工业出版社，2001．

[3] 刘瑛，等．数控技术应用专业英语．北京：高等教育出版社，2004．

[4] 尹小莹，杨润辉．外贸英语函电—商务英语应用文写作．西安：西安交通大学出版社，2008．

[5] 谢小苑．科技英语翻译技巧与实践．北京：国防工业出版社，2008．

[6] 张谡．实用业务英语．北京：外文出版社，2003．

[7] 赵运才，何法江．机电工程专业英语．北京：北京大学出版社，2006．

[8] 翟天利．科技英语阅读与翻译实用教程．北京：新时代出版社，2003．

[9] 綦战朝．机电专业英语．北京：清华大学出版社，2007．

[10] 杨成．机械专业职场英语．北京：电子工业出版社，2009．

[11] Kennis. Reengineering Australia forum.australia, 2006.

[12] Lindberg R A. Processes and Materials of Manufacture.1983.

[13] Mallow R A. Plastic Part Design for Injection Molding. Cincinnati: Hanser/Gardner Publications, Inc., 1994.

[14] Jameson E C.Electrical Discharge Machining Tooling, Methods and Applications.1983.

[15] Polywka J. Programming of Computer Numerically Controlled Machines. New York: Industrial Press, 1992.

[16] Wright P K. 21 世纪制造．北京：清华大学出版社，2004．

[17] Rao J S .Mechanism and Machine Theory. New York: Wiley, 1989.

[18] Peter R. Mechanical Design.London: Arnold, 1998.

[19] Yusuf Altintas. Manufacturing Automation-Metal Cutting Mechanics. Machine Tool Vibration and CNC Design. UK; Cambridge University Press, 2000.